giving
nature
a home

RSPB
BRITISH
NATUREFINDER

Marianne Taylor

BLOOMSBURY WILDLIFE

LONDON · OXFORD · NEW YORK · NEW DELHI · SYDNEY

BLOOMSBURY WILDLIFE
Bloomsbury Publishing Plc
50 Bedford Square, London, WC1B 3DP, UK

BLOOMSBURY, BLOOMSBURY WILDLIFE and the Diana logo are trademarks of
Bloomsbury Publishing Plc

First published in Great Britain 2018

A catalogue record for this book is available from the British Library

Library of Congress Cataloguing-in-Publication data has been applied for

ISBN: PB: 978-1-4729-5127-4; ePub: 978-1-4729-5126-7; ePDF: 978-1-4729-5125-0

2 4 6 8 10 9 7 5 3 1

Designed by Rod Teasdale

Printed and bound in China by C&C Offset Printing Co., Ltd.

To find out more about our authors and books visit www.bloomsbury.com
and sign up for our newsletters

Contents

Introduction

The British Isles abound with wildlife, from the majestic to the minuscule and everything in between. Our land also teems with wildlife watchers, many making their first forays into the natural world, inspired by ever more stunning camera work from the natural-history film-makers and photographers, and eager to witness the wonders for themselves. This book aims to bring wildlife and wildlife watcher together, helping you to see more and to get more out of your sightings.

Which wildlife?

This book is a companion to *RSPB British Birdfinder*, a 'how to find' guide to all the birds that regularly breed in Britain or visit at other times of the year. Now, *RSPB British Naturefinder* offers the same guidance for all the other animal groups found in Britain. Whether you are eager to see your first Otter, observe the mating dance of the Great Crested Newt, admire the dashing beauty of the Swallowtail butterfly, or discover the solitary bees using your garden, you'll find the help you need here.

We cover some animal groups in full in this book – for example, there are accounts for all British land mammals, and all regularly breeding butterflies and Odonata (dragonflies and damselflies). Other groups, such as moths and hoverflies, are too large for this to be possible, so we focus on some of the most sought-after and striking examples, and offer more general guidance for the rest of these groups. As this book is aimed at the beginner and keen amateur, we have concentrated on those animal groups that are easier to find and observe, so for reasons of space some have not been included.

The book focuses on land animals but includes some that are found in freshwater or at sea, provided they can be observed with reasonable ease from land or a boat. So whales and dolphins are included, as are some freshwater and rockpool fish, but marine fish are not, except two species (Basking Shark and Sunfish) that are regularly seen at the surface and may be spotted from land, or ferry crossings or wildlife-watching boat trips.

You may not expect to find a **Fox** on a beach, but wildlife-watching is often unpredictable.

Watching wildlife responsibly

This book takes the starting point that the animal's well-being is paramount, whether that animal is a seal, snake or spider. Many rarer British animals of all kinds have full or partial legal protection, but we should still treat those that do not with the same care. Large mammals will often react strongly and negatively to human presence, so you should always watch them from a distance or a good hide and be particularly careful around young animals. From time to time a mammal will approach you, and this can make for a beautiful encounter but requires patience and care. Chasing after any animal isn't likely to result in better views and may severely stress the animal.

For some particularly sensitive or elusive species, we recommend you seek them out through organised guided walks, rather than attempt to find them yourself. This not only dramatically improves your chances of a sighting but ensures you will not cause untoward disturbance.

We strongly discourage any attempt to handle wild mammals, reptiles or amphibians, unless the animal is injured and in need of rescue, or needs to be removed from immediate danger. When it comes to familiar and easy-to-handle invertebrates, please use care and common sense. Carefully picking up an insect for a closer look does little harm, but you should never grab one by the body, wings or legs – instead, offer an insect a finger to step onto, or cup your hands around it. When you have finished observing it, gently place it back where you found it. Be aware that some invertebrates can bite or sting; not all of the potential biters are obvious (for example, water boatmen are capable of inflicting a surprisingly painful nip).

As well as treating the animals themselves with care, be mindful of the Countryside Code (see p282) whenever you are out and about in the countryside – close gates, leave no trace of litter and avoid trampling crops or habitats. When you are intent on getting close-up views of one particular species of insect in a field, it is all too easy to forget about all the other living things present and leave a trail of destruction in your wake. Always give yourself as much time as you can, step with care, keep to paths and trails, and move slowly and softly. Not only will these measures help look after wildlife and wild places in general, they will also improve your chances of seeing your target species.

Large dragonflies like this male **Brown Hawker** are usually hyperactive but are more sluggish on cooler days.

Sharing knowledge

In this age of social media, much information on wildlife sightings is passed on via Twitter, special interest groups on Facebook and Flickr, and individual blogs. Sharing sightings is a rewarding way to connect with other people who share your interest. These resources are absolutely invaluable to the brand-new beginner, too. If, for example, you want to learn more about identifying hoverflies or shieldbugs, sign yourself up to a relevant group and you will have free, easy access to a wealth of expert knowledge and experience. This will also put you in touch with other enthusiasts who live in your area, and then meet-ups and group outings to productive local sites can be arranged. (The usual provisos of internet safety apply, of course.)

Not every sighting should immediately be shared with the world, though. Some species are highly sensitive to disturbance, and not every purported wildlife watcher is well-intentioned. There are still those butterfly collectors who would target vulnerable protected species and those who harbour ill-will towards Adders. Persecution of carnivorous mammals is an issue in some areas. If you are in any doubt about the wisdom of spreading word of a sighting, tell no-one except the local recorder for the county where you made the sighting. Each county has its own coordinated recording scheme for wildlife, and most will have a dedicated recorder for mammals, butterflies, moths, dragonflies and damselflies, and often other groups. These links will help you to find local recorders, or offer online forms for you to submit reports, which will then be passed on to the relevant recorder:

Mammals: mammal.org.uk/science-research/surveys/county-mammal-recorders/
Reptiles and amphibians: arc-trust.org/report-your-sightings
Butterflies: butterfly-conservation.org/2390/Recording-contacts
Moths: butterfly-conservation.org/16494/county-moth-recorders
Odonata: british-dragonflies.org.uk/content/county-dragonfly-recorders
Beetles: coleoptera.org.uk/contacts

For other groups, contact your local RSPB office (rspb.org.uk/about-the-rspb/get-in-touch/rspb-offices) or Wildlife Trust (www.wildlifetrusts.org/map) in the first instance, for advice.

At home and away

Exploring the countryside is all part of the joy of watching wildlife and to see many of the particularly coveted species, you'll need to make a long journey. This does, of course, carry costs, both financial and environmental. Minimise the impact of your trips by planning carefully. For example, a well-timed and well-organised week-long summer trip to the west Highlands could allow you to find and observe mammals such as Pine Martens, Otters, Red Squirrels, Red Deers and a range of sea mammals and interesting shoreline and rockpool creatures, as well as rare insects like Large Heath and Chequered Skipper butterflies and Azure Hawker and Northern Emerald dragonflies.

All wildlife watchers should do their bit to protect the environment by using public transport or going by bike or on foot where possible. This is easier said than done in some areas, but it's worth taking the time to investigate options and costs.

Where you can contribute most, though, is in your own local area, and (if you have one) particularly in your garden. Become the warden and wildlife recorder for your own small patch: create an outside space that is attractive to wildlife of all kinds and keep a record of everything that occurs there. In terms of identification, some animal groups are far easier to get to grips with than others but all have the potential to surprise and delight as you discover more and more about them.

How to use this book

Common names and, for individual species, a scientific name are at the top in the heading.

A species' size is given at the top of the information panel alongside the main photograph. Sizes are mostly listed in centimetres, but where a species' size exceeds 2m, it's given in metres, and for sizes under 20mm, it's given in millimetres. In most cases, the measurement provided is the total body length for a large adult (including its tail where relevant). For butterflies and moths, however, the wingspan is given, and for crab, the width is listed. Where a group rather than a single species is covered on a single page, to avoid confusion the size may have been omitted.

Below the size, the habitat types each species generally prefer are shown from these options:

| town and garden | woodland | open lowlands | open uplands | river and lake | marsh and estuary | at sea |

Underneath these symbols is a numerical representation of how common and easily found each species is in Britain on a scale of 1 to 5 with 1 being widespread and common and 5 being very rare and localised. However, these figures have been weighted according to a species' abundance in the more populated parts of Britain.

Beneath that is a calendar showing the best months to look for that species.

In accounts for some of the rare or localised species, a selection of **Super Sites** is included. These places are good for a particular species and are visitor-friendly – many are RSPB reserves or other nature reserves. While these lists cannot be exhaustive, we have tried to select sites that hold a particularly good diversity of other wildlife as well as the species in question, to allow you to enjoy more productive visits.

Under the main photo, a short introduction describes each species' general appearance (including how to separate it from similar species) and its distribution in Britain and Ireland.

The **How to find** advice has been divided into three sections:

The **Timing** section describes which time of year and what time of day to look for a species and, in some cases, the weather conditions under which you are most likely to see it. Where relevant, this section also states which life-cycle stages occur at particular times of the year and day. For insects, Timing describes when the adult insect is active, only mentioning immature stages if these are particularly easy to find.

The **Habitat** section describes where you will see the species along with any specific habitat needs it may have, for example, the larval foodplants that butterflies require.

The **Search tips** section provides more specific advice on how to search for and locate a species, once you are in the right place at the right time.

At the bottom of each species account, the **Watching tips** box offers advice on getting the most out of your sightings. Here we describe ways to enjoy more extended sightings without stressing the animal, flag up interesting behaviours to look out for, offer ideas on how to encourage a species into your garden, and detail how to pass information about less common species to local wildlife recorders.

Interspersed among the species accounts, you will find several **themed page spreads** such as 'Small Mammals' on pages 18-19. These pages deal with the practicalities of various more general aspects of wildlife-watching, and they cover activities such as using bat detectors and moth traps, as well as pond-dipping and rockpooling.

Grey Squirrel

Sciurus carolinensis

Length 40–55cm

| 1 | 2 | 3 | 4 | 5 |

J F M A M J J A S O N D

This North American squirrel was introduced to Britain around the start of the 20th century and has spread and thrived (at the expense of the native Red Squirrel). It is now our most easily observed wild mammal over much of mainland England, Wales and southern Scotland. Intelligent and opportunistic, it is a controversial species – much loved by some, but unpopular for the damage it can do to trees and its skill at raiding garden bird-feeders, as well as its impact on Red Squirrel populations. Grey Squirrels are larger and more robust than Reds, lack ear-tufts and have grey fur (though often with brown tones). A melanistic (black) form is common in some areas, and albinos are occasionally seen.

Albino and melanistic Grey Squirrels tend to occur in localised pockets. This is because these unusual colours are caused by genetic mutations, so concentrate in smaller, interbreeding populations. Albinos are vulnerable to predators and have poor eyesight, so often survive best in urban parks where predators are rare.

How to find

■ **Timing** Grey Squirrels are diurnal, with activity highest in the early mornings and the evenings. They do not hibernate and can be seen all year round, though they are probably at their most active in spring and again in autumn. Kits are born in spring and some females have a second litter in summer.

■ **Habitat** Any area with deciduous woodland – rarer in pine forests. Parks and gardens with mature trees hold the densest populations.

■ **Search tips** These squirrels forage on the ground as well as in trees, and are not at all discreet in their activities. They will come to garden birdfeeders and bird tables. Those living in town parks can become completely fearless. Grey Squirrels can often be located by the scolding alarm call they give when anxious, which sounds like a drawn-out chicken-like cluck.

WATCHING TIPS

It is easy and rewarding to watch Grey Squirrel activity, from their frantic spring courtship chases that take them spiralling into the tree-tops, through drey-building and care of the young to preparing for winter by making food caches. Many people would rather discourage than encourage these squirrels in the garden, but if you do want to attract them then try a squirrel-specific feeder containing peanuts, to keep them away from your birdfeeders. With some imagination you can even set up squirrel obstacle courses or puzzles with food rewards, to study their impressive agility and problem-solving skills.

Sciurus vulgaris

Red Squirrel

Length 35–45cm

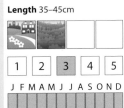

1	2	**3**	4	5

J F M A M J J A S O N D

The only squirrel native to Britain, this species seems unable to coexist with the introduced Grey Squirrel and has disappeared from most of Britain. It can still be found in areas without Greys – northern Scotland, the far north of England, in coastal Merseyside and on some islands, such as the Isle of Wight, Scillies and Brownsea Island, as well as Anglesey where Greys have been eradicated. It is smaller than the Grey Squirrel with rich fox-red or dark red fur (fading to blond on the tail in some individuals), a white underside, and long upright ear-tufts in winter.

In summer, Red Squirrels lack ear-tufts and have a shorter and brighter red coat.

Super sites

★ RSPB Abernethy Forest
★ Brownsea Island, Dorset
★ Formby, Merseyside

How to find

■ **Timing** Red Squirrels are most active early and late in the day. They do not hibernate so can be seen year-round.
■ **Habitat** In Britain, associated with mature pine forest but will use other woodland types in areas where Grey Squirrels are absent. They will also visit garden feeders.
■ **Search tips** Depending how accustomed they are to people, Red Squirrels may be timid or confiding. They are easiest to watch at well-established feeding stations, as can be found at many nature reserves within their range (for example at RSPB Loch Garten, near Boat of Garten in Speyside). If you are staying in Red Squirrel countryside, even just for a short visit, it is worth putting peanuts or mixed seed out in the garden to attract them. When walking in a forest that's home to Red Squirrels, listen for them moving through the branches above. When agitated, Red Squirrels may give a high-pitched wailing alarm call.

WATCHING TIPS

Red Squirrels are very active, charming mammals, which spend more time in the treetops than Greys do, but will come down to ground level in search of food or when engaged in their frenetic springtime courtship chases. Two or more males may chase a female through the trees for up to an hour, before she accepts one as a mate. Red Squirrels are more agile than Greys and are most entertaining to watch as they dangle by their hind feet from branches to reach hanging birdfeeders. Extra food may be buried or hidden in cracks in tree bark for later consumption.

Brown Rat

Rattus norvegicus

Length 35–50cm

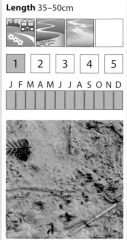

1	2	3	4	5

J F M A M J J A S O N D

The Brown Rat is a large, stout rodent with relatively small eyes and ears, and a long hairless tail. It is native to Asia but has colonised all continents thanks to its habit of stowing away on ships, and its ability to thrive in urban environments. It has been present in Britain since at least the 18th century and is very common, though often shy and difficult to see. It is generally considered a serious pest for its potential to raid food stores, spread diseases and cause structural damage – it can also harm native wildlife, particularly burrow-nesting seabirds. Nevertheless, it is an interesting and intelligent animal – the wild ancestor of domesticated 'fancy rats', whose doting owners would testify to its charms.

Unwelcome though they often are, Brown Rats can be valuable components of an ecosystem, acting as 'clean-up' agents by scavenging carrion and rubbish discarded by people, for example. They are also a valuable food source to predators such as Stoats, Polecats and Foxes, as well as birds of prey.

How to find

■ **Timing** Brown Rats are most active at night but are often out and about in daylight as well. They are active all year round but are probably easier to see in winter, when food sources are scarcer and more concentrated.

■ **Habitat** Any lowland habitat with a reliable food supply may hold Brown Rats, and populations can be very dense in areas with abundant food. Gardens, parks, lake shores and riversides are the likeliest places to see them. They nest in very sheltered or underground spaces, including within buildings.

■ **Search tips** Bird feeding stations in gardens (and on nature reserves) attract rats. They take food placed or spilled on the ground, but can also climb up to hanging feeders. However, a lack of ground vegetation for cover may discourage them. Brown Rats are attracted to areas in town parks where people regularly feed birds, though they rarely become anywhere near as bold as Grey Squirrels. They are strong swimmers and can often be seen crossing open water, leading to confusion with Water Voles.

WATCHING TIPS

Brown Rats are usually quite timid – care and patience is needed to watch them at length. Stay quiet and still, though, and they may approach closely and provide good opportunities to watch their social interactions and their impressive swimming and climbing skills. Most people do not really want to watch rats in their own gardens, but bird feeding stations at nature reserves (particularly those with lakes and some woodland) can offer good viewing opportunities. High numbers of Brown Rats may also mean a higher chance of seeing predators such as Stoats and Polecats.

Rattus rattus

Black Rat

Length 28–40cm

| 1 | 2 | 3 | 4 | 5 |

J F M A M J J A S O N D

This smaller, daintier cousin to the Brown Rat is also an introduced species, native to Asia. However, it has been with us much longer, arriving some 2,000 years ago. Once an abundant and notorious pest, it has been comprehensively outcompeted by the Brown Rat and is now one of our rarest mammals. However, it remains a potentially damaging non-native and has (controversially) been eradicated on some islands, to protect breeding seabirds. It usually has black or dark grey fur, with larger eyes and ears than a Brown Rat, giving a more mouse-like look. It has a proportionately longer tail, and is faster-moving and more agile.

Although it is now extremely rare in Britain, the Black Rat is one of the world's most abundant and widespread mammals. In some urban areas it suffers little or no persecution but is accepted or even welcomed. Karni Mata Temple in Rajasthan, India, is home to 25,000 'holy' Black Rats, which are revered by priests and visitors alike.

How to find

■ **Timing** Black Rats are active all year round and may be seen at any time of day, though night-time, dawn and dusk are most likely.

■ **Habitat** In Britain now, Black Rats can only be found on rocky coasts on a few small offshore islands, and occasionally in warehouses and other buildings where food might be found, in port towns.

■ **Search tips** Seeing Black Rats in the wild in Britain today is a near-impossible challenge, following its eradication on Lundy Island and the Shiants (Hebrides). Black Rats may survive on some other Hebridean islets, but there is currently no known site which is easily reachable by visitors where they can be observed with certainty. They are still found in dock areas of port towns such as Liverpool at times, but are most likely to turn up in private property instead, and their presence is no more tolerated than that of their common cousin. If you see a black-furred rat, take note of its body proportions, as melanistic Brown Rats can occur.

WATCHING TIPS

Black Rats can be attracted by food baits, particularly when placed close to cover or shelter. These are shy animals in most situations, and feel more comfortable feeding in places where they can make a quick escape if startled. Watching captive Black Rat colonies at zoos (such as the British Wildlife Centre in Surrey) reveals this is a hyperactive, social, very intelligent and acrobatic little mammal and a joy to observe. Some wildlife watchers have lamented this rat's almost complete disappearance from Britain, but as it is non-native (and can be so destructive), no reintroductions will be planned.

Wood, House and Yellow-necked Mice

Apodemus and *Mus* genera

Length
Wood Mouse 18cm
House Mouse 16cm
Yellow-necked Mouse 22cm

1	2	3	4	5

J F M A M J J A S O N D

Wood Mouse / House Mouse

Wood and Yellow-necked Mice are closely related and very similar. Both have large, prominent eyes and ears, warm brown fur (white on the belly), and a tail about as long as their bodies. The House Mouse has a proportionately shorter tail and smaller eyes and ears. The Yellow-necked Mouse has a yellow upper chest and throat – the Wood Mouse may have a small yellow spot but never a full band across the chest. Wood and House Mice are extremely common and widespread, while the Yellow-necked Mouse is more localised, occurring mainly in the South-East and East Anglia. All are primarily nocturnal, timid and shy.

The Yellow-necked Mouse, though rarer, is more likely to be found inside homes than the Wood Mouse. All three of these mouse species may come into houses, especially in winter, though only the House Mouse is likely to live permanently in occupied buildings.

How to find

■ **Timing** Mice do not hibernate but will be much less active in cold weather. Late summer and autumn is the easiest time to see them, as they are collecting extra food for winter larders (also the populations are highest). The best times of day to see them are dawn, dusk and night-time.

■ **Habitat** Lowland deciduous woodland and wood edges are probably the best habitats for Wood and Yellow-necked Mice. All species will enter buildings, especially House Mice, and will nest and shelter in bird nestboxes.

■ **Search tips** These mice use regular 'runs' through vegetation, and placing bait in or near such runs gives a good chance of sightings. Using a sand trap for footprints will reveal their presence. Droppings and damage to stored food indicate their presence inside homes. They will visit garden birdfeeders and easily climb to reach hanging bird-feeders – also, make sure your garden has plenty of natural food (native berry- and nut-bearing shrubs), and shelter such as log piles and drystone walls.

WATCHING TIPS

For an intimate insight into mouse life, try placing a hole-fronted bird nestbox with a built-in camera low on a tree trunk, within sheltering vegetation. The hole can face the tree trunk as long as there's space for the mouse to squeeze in. Mice will take nuts and seeds put out on bird tables or the ground. They are timid, so ideally watch from indoors or a hide. House Mice can be very problematic inside homes so it's wise not to encourage this species.

Micromys minutus

Harvest Mouse

Length 12cm

| 1 | 2 | 3 | 4 | 5 |

J F M A M J J A S O N D

This diminutive mouse is Europe's smallest rodent. Its golden fur and relatively small eyes and ears give it an almost hamster-like look, belied by the long prehensile tail that assists it as it climbs among long grass stalks and other vegetation. The Harvest Mouse is fairly common but not easy to see, and agricultural change over the last century has led to some significant declines. Anyone with a largish garden in fairly open countryside could encourage the species by having wild corners with long, tussocky grass and other long-stemmed plants, among which the mice can nest.

Harvest Mice are very popular subjects for photographers, but most of the stunning images you'll see are of captive-bred animals, which can easily be persuaded to clamber about and pose on photographic 'sets'. Photos taken in the wild are much rarer but arguably much more satisfying.

How to find

■ **Timing** Harvest Mice are active at dawn and dusk as well as at night, and may be out more in the daytime during winter than in summer.

■ **Habitat** Open, well-grown meadowland and crop fields are home to this species, and they may also be found at reedbed edges. Farmland managed in a reasonably natural way, with varied habitats and plenty of wild patches, is most likely to support them.

■ **Search tips** This is a very alert animal and will vanish at the slightest disturbance. The best chance of a sighting is to locate a nest. These are round, tennis ball-sized, tight bundles of woven grass and other material, suspended half a metre or more above ground level within the vegetation. Because they rarely descend to ground level you stand little chance of finding their tracks, but they may come to bait.

WATCHING TIPS

You will need great luck and patience to watch wild Harvest Mice, and the nature of their habitat does mean sightings will probably be brief – but it is still well worth the effort to see this delightful little animal. Nests are much easier to find in winter, but if you return to the same area on a summer evening you may be lucky enough to see juvenile mice, which will be a little less nervous than the adults. Some observers have had success by placing half a tennis ball, filled with bait, among the grass stems a good metre or so above ground level.

Bank and Short-tailed Voles

Myodes glareolus,
Microtus agrestis

Bank Vole   Short-tailed Vole

Length
Bank Vole 15cm
Short-tailed Vole 16cm

1	2	3	4	5

J F M A M J J A S O N D

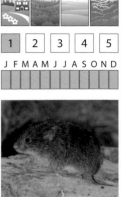

These two voles are very common and quite similar at first glance. The Bank Vole has a more mouse-like appearance with larger eyes and ears, a longer tail and more rufous fur. Voles are easily told from mice given a good view, by their small furry ears that are mostly hidden by the fur on the head, and the relatively small eyes and short tails. Mice have large, bare ears, big bulging eyes and very long tails.

Although much more abundant than the Bank Vole, the Short-tailed Vole is more difficult to see. This is because it spends most of its time shuttling along its 'runs', close to ground level in long grass, while Bank Voles will explore more open ground and climb in bushes.

How to find

■ **Timing** Both voles are active all year round and at all times of day, though activity is highest between dusk and dawn.
■ **Habitat** Bank Voles live in woodland, hedgerows and other habitats with trees and shrubs, including gardens. Short-tailed Voles prefer open habitats, particularly rough grassland but also heaths and moorland.
■ **Search tips** Although the rarer species, the Bank Vole is more often seen, as it visits bird feeding stations. It can climb but less ably than mice. Exposed, tangled tree or shrub root systems are good places to watch for it – look for the holes to its burrows. The Short-tailed Vole is best searched for by finding its runs at ground level in long grass. Exposing a short stretch of a run and baiting it may produce sightings.

WATCHING TIPS

As with other small rodents, voles are easiest to see if attracted with bait such as nuts and berries, but if you find a Bank Vole's burrow system and watch quietly, you could have longer views of the animals coming and going, foraging and interacting. The same goes for Short-tailed Voles in their runs. Both voles have poor eyesight so can be watched at close range without a hide as long as you keep quiet. Fields with an abundance of Short-tailed Voles will attract predators like Weasels and Stoats, and also (particularly in winter) birds of prey like Barn and Short-eared Owls and Hen Harriers.

Arvicola amphibius

Water Vole

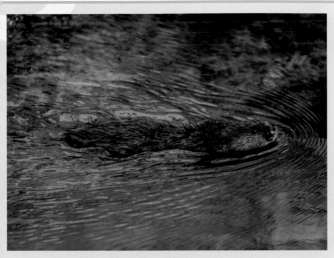

Length 28cm

| 1 | 2 | 3 | 4 | 5 |

J F M A M J J A S O N D

The Water Vole is our largest vole and is specialised for an aquatic lifestyle, rarely venturing far from the waterside. Its numbers plummeted following the introduction of the North American Mink, but it is rallying in some areas, and has been successfully reintroduced to many nature reserves. Its blunt nose, small eyes and ears, and short tail separate it from the Brown Rat. It is easily told from other voles, being much larger and bulkier, with very small eyes and ears, and is rarely found away from the waterside.

With its hefty body and almost invisible ears, the Water Vole is easily identified.

Super sites

★ RSPB Rainham Marshes, Essex
★ RSPB Rye Meads, Hertfordshire
★ RSPB Saltholme, Cleveland
★ RSPB Strumpshaw Fen, Norfolk

How to find

■ **Timing** Water Voles do not hibernate as such, but are more difficult to see in winter as they stay near their nests, living on stored food. They are active by day and night.

■ **Habitat** Slow-flowing rivers, marshes, ditches and lakes with a good amount of marginal vegetation.

■ **Search tips** Water Voles leave several signs of their presence, including burrows in the bankside, with cropped grass at the entrance, latrines with piles of cigar-shaped 1–1.2cm-long droppings, and food remains in piles (nibbled reed and sedge stems, with 45-degree cuts at their end). When disturbed, they drop into the water with a definite 'plop'. Watch for voles swimming across channels or close to the shore. Slight movement in bankside vegetation may reveal a feeding vole. They may come to bait, such as quartered apples.

WATCHING TIPS

Like other voles, Water Voles are short-sighted and can be watched easily if you keep still and quiet. Wait on a bridge over a reed-fringed canal or stream and watch them swimming along, towing chunks of reed stem along with them, or sitting close to the water's edge. Listen, too, for signs of activity among waterside reeds or sedges - you might even hear the crunching of their incisors as they nibble at the plants. If you spot an American Mink while watching Water Voles at a nature reserve, be sure to inform the reserve staff, as minks are extremely harmful to Water Vole populations.

Hazel Dormouse

Muscardinus avellanarius

Length 14cm

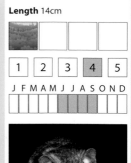

| 1 | 2 | 3 | 4 | 5 |

J F M A M J J A S O N D

This beautiful small mammal is uncommon and difficult to see as it has a very long hibernation period and when active it mostly keeps to the trees, climbing from twig to twig and almost never coming to ground level. It can be told from other small rodents by its long, fully furred tail. It is mainly found in southern England, and parts of Wales. In some areas, aerial bridges have been built to help Hazel Dormice move safely from trees on one side of a road to the other. These dormouse walkways can help isolated populations to reconnect, improving genetic diversity.

The chances of seeing a Hazel Dormouse in daylight are small – with its big eyes, this animal is adapted to a nocturnal lifestyle.

Super sites

- ★ RSPB Tudeley Woods, Kent
- ★ RSPB Broadwater Warren, Kent
- ★ RSPB Swell Wood, Somerset

How to find

■ **Timing** Hazel Dormice are in hibernation from mid-autumn to late spring. When active, they are mainly nocturnal.

■ **Habitat** Found in deciduous woodland and mature, dense hedgerows with hazel, that adjoin areas of woodland. May also occur in large gardens with trees and shrubs.

■ **Search tips** These dormice are usually located by nestbox surveys – they will make winter nests inside bird nestboxes and purpose-built dormouse boxes. However, it is illegal to handle or disturb a dormouse without a special (Schedule 5) licence. The most characteristic food remains are hazelnut shells accessed through a round hole. There are tooth marks on the outside of the hole, and smooth gnaw marks around its circumference.

WATCHING TIPS

The easiest way to see a Hazel Dormouse is to join an organised nestbox check with licensed conservation workers – nature reserves may organise such days. You will be able to see sleeping dormice curled up in their nests, and will learn a great deal about the species and the various conservation projects underway to help encourage and protect it. To see active dormice, watch at dawn or dusk for activity in treetops or high parts of hedgerows, especially later in summer when juveniles disperse to new sites. Binoculars may help you to spot them as they climb nimbly along slender twigs, with much adjustment of tail position to help them balance.

Glis glis

Edible Dormouse

Length 27cm

1	2	3	4	5

J F M A M J J A S O N D

Introduced in the early 20th century, this European species is considered invasive in Britain – it can damage trees, and is a potential crop pest and a damaging competitor with native wildlife. It is also rather predatory for a rodent, and will enter bird nestboxes to take eggs and chicks - it has even been known to prey on roosting bats. However, it is still rare away from its Chilterns stronghold. It resembles a small, thinner-tailed Grey Squirrel, with a smudge of dark fur around its eyes, and a pink nose. It does not habitually sit upright, squirrel-style, and, although it is a skilled climber and leaper, it tends to move more slowly and deliberately than a squirrel does.

Like the Hazel Dormouse, this is a very sleepy animal, hibernating for well over half the year.

Super sites

★ Wendover Woods, Bucks
★ Ashridge estate, Herts/ Bucks

How to find

■ **Timing** Edible Dormice hibernate between October and May, and are nocturnal.

■ **Habitat** This is a woodland species, favouring mature deciduous woodland of oak and beech, but will visit hedgerows and orchards (where it can be damaging to fruit crops). It may also enter buildings, especially loft spaces, and can cause damage by chewing wiring.

■ **Search tips** This dormouse is mainly arboreal. It feeds on tree buds and chews at bark on twig tips, and its presence may be revealed by dead or dying tree-tops, with frayed bark strips dangling from the twigs. It may nest or sleep in larger bird nestboxes. When active at night, Edible Dormice give frequent squeaking calls, and are noisy generally with much snuffling.

WATCHING TIPS

Being nocturnal and a tree-dweller, the Edible Dormouse is difficult to see, though may easily be heard. Head out at night in areas that show signs of its presence, and use a torch to light up trees when you hear activity. If Edible Dormice set up home in your loft you could set up cameras to watch them, but beware their propensity to gnaw at wiring. If you know you have these dormice living locally, be aware that they could attack bird nestboxes, and may even kill adult birds as well as taking eggs and young.

Small Mammals

Mice, voles and shrews vastly outnumber larger mammals, but their size and habits make them difficult to see. They are prey to all manner of other animals, from raptors and snakes to carnivorous mammals like Weasels that are only a little larger than themselves. They can only hope to survive by being discreet in their habits, very good at concealing themselves and able to make a quick getaway to a safe hiding place whenever danger threatens. To find, see and watch small mammals properly takes care and patience, as well as some insider knowledge and a few clever tricks.

Tracks and signs

When moving at ground level, these mammals tend to get about via 'runs' through grass or leaf litter – effectively the runs are tunnels that keep them concealed from above. They scent-mark their runs, with urine or glandular secretions, to advertise their presence to others of their species. You can see whether a run is in use by placing a 'sand trap' in or close to it, and checking it for footprints and tail drag marks. At its simplest, this is just a patch of soft sand. Make things more sophisticated by making a small, shallow tray with a measuring scale on its sides, then filling with sand before placing in the run, or close

House Mice are never welcome house guests. Prevent access by blocking off even the tiniest holes.

to baits. Using the scale will help you determine the species visiting. A basic guide to footprint identification can be found at: ypte.org.uk/factsheets/wildlife-id/footprints-mammals, while books such as *RSPB Nature Tracker's Handbook* have more detailed information. Keep in mind that all mammals' tracks can vary in appearance, with age, sex, and the sort of ground on which they fall (for example, on very soft ground tracks are more splayed out than on firmer ground).

Food remains and droppings also reveal the presence of small mammals. Grey Squirrels, Wood Mice, Bank Voles and Hazel Dormice all enjoy hazelnuts and open them in their own particular ways. For Water Voles, chewed sections of reed stem, and latrines full of droppings are both easily found. Shrews leave behind the harder parts of their prey – for example, discarded caddisfly cases on a stream bank indicate the presence of Water Shrews.

Placing bait

Most of our small mammals forage actively all year round, though they also build food stores, particularly as winter approaches. They will readily take food put out for them, though their presence in the first place will be dictated by good supplies of natural food. Offer nuts and berries for mice and voles, and mealworms for shrews, placed on a level surface at, or near, ground level or within easy climbing reach, and close to cover. Observe quietly, perhaps from a hide, for the best chance of watching their activities at length. Bait may attract less desirable visitors, such as Brown Rats and House Mice, even if placed well above ground level. Be ready to discontinue use if you see these species.

Domestic cats prey heavily on small mammals. Keep yours in from dusk to dawn to help protect wildlife.

Shrews are easily told from rodents by their long, delicate snouts and pinprick eyes.

Spotlighting

Nocturnal mammals, both large and small, are easiest to find if you use a powerful spotlight with a red or a green filter. Be aware, though, that this should never be done in the vicinity of roads or livestock, nor anywhere near a bat roost.

Live trapping

Various humane traps exist which will allow you to catch small mammals and get a close look at them before you release them. This will help you determine which species are present in your area, though it won't help you to observe natural behaviour, of course. The most widely used is the Longworth Small Mammal Trap, which you can buy from suppliers such as NHBS (nhbs.com). Note that a licence is needed to trap shrews – a version of the Longworth trap exists with a 'shrew hole', allowing any caught shrews to escape while keeping in mice and voles, which are slightly larger. Provision traps with plenty of food and nesting material, check them frequently, and if it's necessary to handle caught mammals then do so with great care – they are fragile, and they can and will bite.

Baby squirrels may be very approachable when they are fresh out of the nest and still reliant on parental care.

Beaver

Castor fiber

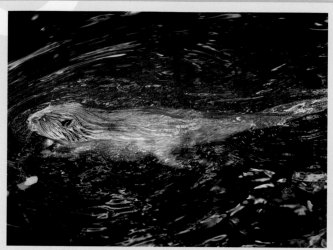

Length 130cm

1	2	3	4	5

J F M A M J J A S O N D

Native to Britain, but eradicated in the 16th century, the Beaver has now been reintroduced in Argyll and Devon, with other projects planned elsewhere. It is an important species ecologically because of the ways it modifies river flow. Its presence is linked to improved river and woodland biodiversity and a reduced risk of flooding. It is a very large, heavyweight rodent which leaves unmistakeable signs of its presence in the form of lodges and dams. Seen clearly, it cannot be confused with any other swimming mammal.

The return of wild Beavers to our countryside could potentially transform riverine landscapes for the better. In years to come, it should be possible to see them at several other sites.

Super sites

★ Knapdale Forest, Argyll
★ River Otter, Devon

How to find

■ **Timing** Beavers are active all year round, though they mainly use stored food in the winter and do not forage so widely. They are mainly nocturnal.

■ **Habitat** Beavers live alongside rivers and ponds in wooded habitats.

■ **Search tips** Signs of Beaver activity are very obvious. Beavers fell trees to obtain building materials for lodges and dams. Dams are large piles of sticks, chewed through and arranged horizontally across flowing water. The damming creates still ponds, in which the Beavers build their nesting 'homes' or lodges. A Beaver lodge can be several metres across and is a low dome of sticks. Lodges are mainly occupied in winter and steam visibly on cold days. Beavers may also create channels or canals of still water near their dams for easier transport of sticks.

WATCHING TIPS

Watching quietly downwind of lodges and dams may produce sightings, especially at dusk and dawn, and particularly in autumn when Beavers are renovating their lodges and bringing in food ready for the winter. You will need patience though, so wrap up warm for a long wait. In summer they wander more widely. Your local Wildlife Trust is likely to be interested in all observations, as Beavers continue to spread along river systems.

Oryctolagus cuniculus

Rabbit

Length 50cm

1	2	3	4	5

J F M A M J J A S O N D

Rabbits were introduced to Britain from mainland Europe in Roman times and have spread successfully throughout Britain, to the point of being a crop pest in some areas. They are easy to find, identify and observe. They are smaller, shorter-eared and shorter-legged than either the Brown Hare or Irish Hare, with dark eyes. The vast majority of Rabbits in Britain have brown 'agouti' coats, but in some areas black Rabbits are quite frequent, and other colours such as ginger and white may occur - usually the result of pet domestic Rabbits entering the gene pool.

Although they are very common, Rabbits tend to be skittish and difficult to approach, but it is worth taking the time to habituate them to your presence. They exhibit a complex and sophisticated range of social behaviours, and are fascinating to watch, especially in spring and summer when breeding activity is at its peak.

How to find

■ **Timing** Rabbits may be active at any time, but are most likely to be seen at dawn and dusk. In some areas that attract a lot of human visitors, Rabbits all but disappear in the daylight hours.

■ **Habitat** Any kind of terrain with soft sloped ground for digging a warren, and grassland for food, can be home to rabbits. Those living in places much visited by people, such as parkland, are often very approachable, but this is not the case anywhere where dogs are regularly walked off the lead.

■ **Search tips** A Rabbit warren comprises a number of burrows (with entrances 10–20cm across) that are close together and dug in sloping soft, often sandy soil. The small (1cm-long) round pellet droppings will be everywhere, and plenty of droppings on well-grazed grassland indicate an area well used by rabbits.

WATCHING TIPS

Rabbits are delightful to watch, particularly through late spring and summer when the kits are leaving the burrows. On a sunny evening, station yourself within sight of a warren with the wind blowing towards you so they are not alerted by your scent, and watch through binoculars. Rabbits have complex dominance hierarchies and the adults' social interactions are interesting to observe, while youngsters seeing the light for the first time are hesitant at first and then exuberantly playful. Rabbit warrens may also attract predators, such as Stoats and Polecats.

Brown Hare

Lepus europaeus

Length 80cm

1	2	3	4	5

J F M A M J J A S O N D

When hormones take over, hares' natural vigilance becomes less acute, and they may even approach quite closely if you keep very still.

Much larger and leggier than a Rabbit, the Brown Hare almost gives the impression of a gazelle with its very long legs. The long, black-tipped ears and golden eyes are also distinctive. Like the Rabbit, this species is non-native, introduced for hunting, but is now widespread, although it has declined enormously over the last century. It is not sociable in the same way that Rabbits are, with no colonial breeding behaviour. However, several may share the same field when grazing, and groups also come together to engage in courtship and mating activity.

Super sites

★ RSPB Havergate Island, Suffolk
★ RSPB Otmoor, Oxon.
★ Cley Marshes, Norfolk

How to find

■ **Timing** Brown Hares are mainly nocturnal but may be seen at any time of day as they rest in the open, in a hiding place called a form, never using any kind of underground burrow. Breeding and courtship activity is at its peak in spring. They can be harder to see in high summer when vegetation is high.
■ **Habitat** Open grassy fields, arable farmland, and heaths and moors are all home to Brown Hares.
■ **Search tips** Dawn and dusk are good times to seek Brown Hares. Scan across fields and along hedgerows, from a high vantage point if possible to spot animals lying down in their forms – look for black ear-tips. Droppings are like those of the Rabbit but about 1.5cm long. Flattened hare-sized patches of grass may be forms.

WATCHING TIPS

There are few wildlife experiences more exhilarating than watching 'mad March hares' in their high-speed courtship chases. Males pursue females to demonstrate their fitness, and try to chase rival males away. Boxing matches occur when a female rebuffs a male's advances. Leverets spend most of their time hidden quietly away separately, but when their mother arrives for the daily feed they often have a brief playtime session, with wild chases and leaps. Hares are usually timid and very alert so watch with binoculars from a good distance – a car makes a good hide.

Lepus timidus

Mountain Hare

Length 70cm

| 1 | 2 | 3 | 4 | 5 |

J F M A M J J A S O N D

This hare is native to Britain, but now restricted mainly to the high uplands of Scotland as it is outcompeted elsewhere by the introduced Brown Hare. It is smaller and shorter-eared than the Brown Hare, and is famous for its coat colour changes, becoming white for camouflage in winter. The Irish Hare is a subspecies found only in Ireland, and it does not moult to a white winter coat, remaining brown year-round. However, RSPB Rathlin Island is home to a number of 'blonde' Irish Hares – a pale colour variant with a gingery coat and blue eyes.

You'll need careful preparation and warm layers for an expedition to see Mountain Hares in their white winter coats.

Super sites

★ Cairngorms, Scottish Highlands
★ Dark Peak area, Peak District
★ RSPB Rathlin Island, Northern Ireland

How to find

■ **Timing** Mountain Hares can be seen year-round, and are fully white-furred between December and early March. They are most active early and late in the day.

■ **Habitat** They inhabit open, rugged, rocky and treeless ground at high altitudes, feeding on heather and other upland plant-life. Irish Hares use all kinds of open grassy habitats.

■ **Search tips** Mountain Hares are well-camouflaged at all times of year and hard to spot. Scan slowly and carefully along suitable slopes with binoculars, checking sheltered spots among boulders. You may flush out hares when walking in the mountains – track their 'flight path' carefully and you may then be able to watch them. In some areas Mountain Hares are quite confiding. Droppings reveal their presence – these are intermediate in size between those of the Rabbit and Brown Hare.

WATCHING TIPS

Getting close to Mountain Hares often means a challenging climb on difficult terrain, but it is worth the effort. They do move down the mountainsides somewhat in winter. Dress warmly and watch with binoculars – like Brown Hares, they begin to chase and box in late winter. Many wildlife tour companies and guides offer specific trips to watch and photograph Mountain Hares – a good (if expensive) way to guarantee close views. Irish Hares are widespread in all kinds of open country in Ireland, and can be quite approachable.

Roe Deer

Capreolus capreolus

Length 120cm

1	**2**	3	4	5

J F M A M J J A S O N D

This small, elegant deer is native to Britain and is quite common and widespread, but shy, and is usually seen in small groups or alone. It is easily recognised by its black muzzle marking, round white rump patch with no obvious tail, and the male's small, upright antlers. In summer its coat is a rich burnished chestnut colour, but it becomes rather greyer and duller in winter. In late winter and spring, when the bucks are growing their new antlers, the antlers look very thick and rounded because of their layer of 'velvet' – this soft skin dries out and wears away from April, leaving the antlers bare with sharp points. They are shed in late autumn and new antlers start to grow immediately.

You may sometimes see Roe Deer feeding in the same fields as cattle. However, the deer seem less keen on the company of sheep, even avoiding sheep-grazed fields for some time after the sheep have been moved away. The interaction and competition between wild grazing mammals and livestock is complex, and varies from habitat to habitat.

How to find

▪ **Timing** Roe Deer can be seen at any time but tend to keep to the shelter of woodlands in the daytime, so are easiest to find at dawn and dusk.

▪ **Habitat** It is found in woodland of all kinds, and adjoining grassy fields (including those occupied by cattle or horses).

▪ **Search tips** A slow drive or quiet walk at dawn or dusk through suitable habitat gives you a good chance of seeing Roe Deer. Scan along woodland edges, and when you spot deer approach slowly, downwind of them. Also listen for the bucks' harsh barking calls, especially in late summer (their rutting season). At this time you may find 'roe rings' – flattened vegetation in circles around trees where bucks have chased does. Roe Deer droppings are shiny black oval pellets, about 1.4cm long. Their slot tracks are narrow, pointed at the ends, about 5cm x 4cm.

WATCHING TIPS

Roe Deer are usually very timid and best watched from a good distance, or a hide. They generally leave the woods to graze in fields in the last hour or two of daylight. Rutting behaviour is hard to observe as it usually takes place among the trees. However, a canopy hide in woodland can give good views. Arrive early at your watching point and stay until the animals have moved on. If you find a young fawn laid up in cover, always leave it well alone, as its mother will return, but your attention could attract a predator.

Cervus nippon

Sika Deer

Length 150cm

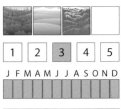

1	2	3	4	5

J F M A M J J A S O N D

Introduced from Japan in the early 1900s, the Sika Deer now has strong populations in Dorset, the New Forest, north-west England and much of Scotland (where it may mix and hybridise with Red Deer). It has a spotted coat in summer (darker than most Fallows), but stags' antlers resemble those of Red Deer rather than Fallow. It is less gregarious than Red or Fallow Deer. Some deer parks have Sikas that are very tame and approachable – watching them is a good way to familiarise yourself with the species before seeking it in the wild.

This deer has an extensive natural range in east Asia, but all Sikas in Britain are originally of Japanese origin.

Super sites

★ RSPB Arne, Dorset
★ Brownsea Island, Dorset
★ Lundy Island, Somerset
★ Forest of Bowland, Lancs

How to find

▍ **Timing** The Sika Deer can be seen at any time of day and any time of year. Its rut takes place in October, and its calves are born in early summer.

▍ **Habitat** Mature woodland (both coniferous and deciduous), young pine plantations, and adjacent heathland and moorland. It is also kept in some deer parks.

▍ **Search tips** Scan suitable countryside from a good distance, checking along the tree line. At rutting time, Sika stags produce a whistling call (Red and Fallow both roar or grunt) and may mark trees with their antlers and break up ground vegetation. Sikas' tracks and droppings are very like those of Fallow Deer – slots are 5–8cm long, 3–5cm wide, droppings are shiny black, round with a slight point at one end and indented at the other, about 1.5cm long.

WATCHING TIPS

Sikas are most interesting to watch in the rutting season, when stags defend a territory and will posture and compete to assemble a harem of females. Their range of vocalisations is impressive, with barks, whistles and even screams. At other times of year, Sikas are quiet, discreet and rather solitary, although you may find them in small groups in winter. From midsummer you can see small calves with their mothers. In areas where Sika and Red Deer both occur, look out for hybrids. These can be variable in appearance but will show traits of both parent species.

Fallow Deer

Dama dama

Length 155cm

1	2	3	4	5

J F M A M J J A S O N D

Fallow does are much smaller and daintier than the mature bucks.

This is our most variable deer species, with white forms relatively common.

This attractive deer was introduced from Europe around 1100, and has spread widely throughout Britain. It is the commonest deer in England, and is also widespread in Wales, Scotland and Ireland. Spotted in summer and plain brown in winter, it is variable with white and melanistic forms frequent, and has a long tail that it constantly flicks. Mature bucks have flattened, 'palmate' antlers, while youngsters ('prickets') show just simple spikes. Fallow Deer live in small or large, sex-segregated herds outside rutting season. They are commonly kept in deer parks, where the full range of colour variation can often be seen.

How to find

■ **Timing** Fallow Deer are active at all times of day and year, but are most likely to be seen in the open around dawn and dusk. Their rut takes place in October, and fawns are born in mid-summer.

■ **Habitat** The best habitat is deciduous woodland with adjacent quiet meadows for grazing. They will also visit farmland, and are widely kept in deer parks.

■ **Search tips** Look out for herds feeding on woodland edges at dusk and dawn. Wild Fallows are shy, slipping back into the woodland when disturbed. Bucks in the rut make far-carrying groans and grunts, and leave hollows in vegetation at spots where they lie down to rest between encounters with rivals. Their tracks are 5–8cm long, 3–5cm wide, and their 1.5cm-long droppings are shiny black, round with a slight point at one end and indented at the other.

WATCHING TIPS

As with other deer, watch from a hide or car for the best chance of longer, closer views. In deer parks, though, they are approachable, to the point where you must take care not to get too close when watching rutting bucks. Behaviour such as parallel-walking and antler-clashing battles is most likely where there are high numbers of mature bucks and competition is fierce. Younger bucks may play-fight at other times of year while in their bachelor herds.

Muntiacus reevesi

Muntjac

Length 83cm

| 1 | 2 | 3 | 4 | 5 |

J F M A M J J A S O N D

Another east Asian deer introduced to Britain, the Muntjac or Reeves's Muntjac is common and widespread in southern England, and is probably the deer most likely to be seen in gardens – it can also be quite damaging. However, it is shy, solitary and very small, so is not particularly easy to see. Its size and short legs make it unmistakeable – it is perhaps more likely to be confused with a stray dog than any other species of deer. Bucks have short, pointed antlers growing from long furry pedicles.

Muntjacs will sometimes sit down on their haunches, which increases their resemblence to a stocky pet dog. Their size allows them to slip quickly away through even quite dense woodland understorey growth.

How to find

▪ **Timing** Muntjacs can be seen at any time of day, especially dawn and dusk. Being native to near-tropical areas, their breeding behaviour has no particular seasonality and fawns may be born at any time of year.

▪ **Habitat** They favour lowland deciduous woodland with a thick understorey. They will also visit large 'wild' gardens, woodland edges with dense hedgerows, and riversides, and may investigate woodland picnic sites and bird-feeding stations for dropped food.

▪ **Search tips** When walking in suitable habitat, you may well have brief encounters with fleeing Muntjacs. Return to the same spot early or late in the day and watch quietly. Look out for their tiny (3cm) cloven tracks on soft ground to find frequently used pathways and station yourself downwind, perhaps in a hide.

WATCHING TIPS

Muntjac does are ready to mate again shortly after giving birth, so will be pregnant almost constantly and can have a small fawn with them at any time. Bucks defend territories, within which several does live, not associating with each other. Their ability to vanish into dense cover makes it difficult to watch their behaviour at length, though if they visit your garden you will have more options for controlling the spaces they use (possibly at the expense of some favourite garden plants).

Red Deer

Cervus elaphus

Length 2.2m

| 1 | 2 | 3 | 4 | 5 |

J F M A M J J A S O N D

Britain's largest deer is a powerful, imposing animal – the iconic 'monarch of the glen'. Mature stags have impressive, spreading, many-branched antlers. Adults have plain reddish coats, while the calves' coats are spotted with white. Red Deer are native to Britain and very common in the Scottish uplands and some Scottish islands, with more scattered populations in England (particularly the South-West) and Wales. They are also widely kept in deer parks. Red Deer are often easy to watch as they tend to prefer open landscapes and are usually seen in sizeable herds. Where they share habitat with more-or-less free-ranging sheep, the two species tend to keep their distance.

Although we associate Red Deer most closely with open moorland, they can be found in a wide range of other habitats as well. Their large size makes them easy to watch even from a long distance away.

How to find

■ **Timing** Red Deer are active year-round and through the day, though in some places retreat to wooded areas in the daytime, emerging to graze at dusk. Their rut begins in mid-September, and calves are born from late spring.
■ **Habitat** Most populations can be found on open grassy and heather moorland. A few live exclusively in woodland. They are also found on heathy lowlands and (in parts of East Anglia) marshland.
■ **Search tips** Red Deer feeding in the open are noticeable from a distance because of their size, and the fact that they are often in large groups. In Scotland, driving over open moorland early or late in the day will almost always produce sightings. The tracks are large (up to 8cm x 6cm), the droppings about 2.5cm long and acorn-shaped. Rutting stags give loud bellowing roars that carry a long distance.

WATCHING TIPS

The Red Deer rut is dramatic, with rival stags regularly clashing antlers to impress the hinds and attract a harem. If you watch the rut in a deer park, keep a good distance – the stags can be quite fearless and very dangerous when awash with hormones. After the rut, the sexes separate again, the males forming loose bachelor groups and the hinds sticking to extended family parties. These can be most interesting to observe, as the dominant hind guides activity and maintains discipline. Through summer, most hinds will have a spotted calf with them.

Hydropotes inermis

Chinese Water Deer

Length 90cm

1	2	3	4	5

J F M A M J J A S O N D

This introduced species is close to Roe Deer in size but looks bulky and small-headed. It has light gingery fur, and the male lacks antlers but has long tusks. It is the most localised of our deer, mainly restricted to East Anglia. While it continues to spread, it does not seem to be increasing in numbers, as although new populations are appearing, others have been lost. It is seldom problematic to farmers. Shy and usually solitary, it can be rather difficult to observe.

The male Chinese Water Deer's long pointed tusks are used in rutting battles, and can inflict serious injuries.

Super sites

★ RSPB Strumpshaw Fen, Norfolk
★ RSPB Minsmere, Suffolk
★ Hickling Broad, Norfolk

How to find

■ **Timing** Like other deer, this species is most likely to be seen in the open around dawn and particularly dusk, and is easier to spot in winter when vegetation growth is lower. The rut is in December, and fawns are born the following May or June.

■ **Habitat** This deer is most often seen on wet, marshy grasslands, and it swims readily. It will also visit farmland and may be encountered in wooded areas close to suitable grassland.

■ **Search tips** Scanning marshland from an elevated position, such as from the tower hides that you'll find at many wetland nature reserves, is a good way to locate Chinese Water Deer – this will also reveal their pathways through the grass. At ground level, both the deer and their runs are difficult to see among lush vegetation. The tracks and droppings are a little smaller than those of Roe Deer. Rutting males give whistling and squeaking calls.

WATCHING TIPS

You may see small groups of Chinese Water Deer feeding together, but this is generally a solitary animal. In summer, lush, marshy vegetation offers concealment for the grazing deer, and often all you will see is a head (adorned with comically oversized Mickey Mouse ears) lifting up from the grass from time to time. Males are fiercely aggressive to each other (many bear scars from rutting injuries). Only where densities are high are you likely to see such encounters. Does often produce twins, triplets or even quadruplets, and can be watched with their young through summer and autumn.

Wild Boar

Sus scrofa

Length 180cm

| 1 | 2 | 3 | 4 | 5 |

J F M A M J J A S O N D

Wild Boar piglets are charming, but their mothers are ferociously protective so give them plenty of space.

The Wild Boar is native to Britain, but was hunted to extinction in the 17th century. Now, this shaggy-coated wild ancestor of domestic pigs is back, following escapes from farms – Wild Boar is a popular gourmet meat. It is established in woodland in a few regions of southern England, most notably the Forest of Dean. Some Wild Boars in Britain show definite signs of mixed ancestry with feral domestic pigs, such as a paler-than-usual coat. Very young piglets have caramel-coloured coats with paler stripes, leading to their nickname of 'humbugs'. However, they develop an adult-like dark coat well before they are anywhere near adult size.

Super sites

★ Forest of Dean, Gloucestershire
★ Beckley Forest, East Sussex

How to find

■ **Timing** Wild Boars are most active at dawn and dusk. They will be out and about all year round and in summer will have small piglets.

■ **Habitat** Their preferred habitat is deciduous woodland with a decent understorey, and they may also visit fields and farmland adjacent to the woods.

■ **Search tips** Signs of boar presence are quite evident. The animals grub up roots to eat, and areas that have been worked over show dug-up patches and a muddle of tracks in the exposed soil. There may also be damage to fencing. They also rest in wallows (muddy pools) and nearby tree trunks may be smeared with mud, and perhaps have scratch marks from tusks. The tracks show four indentations – the deer-like cleats at the front, and two round marks left by the dew claws behind. Droppings are sausage-shaped and about 10cm long.

WATCHING TIPS

When you have found signs of boar activity, see if you can locate the routes used by the animals and watch from a good distance at dusk or dawn. This is one of the very few wild mammals in Britain that poses a real danger to the watcher, so must be treated with respect. In particular, never approach a sow with piglets. Sitting and waiting is usually a good strategy but, if you are not in a hide, be prepared to back carefully away if it looks as though a group is wandering too close to your position.

Length 38cm

| 1 | 2 | 3 | 4 | 5 |

J F M A M J J A S O N D

This charming small predator is common throughout Britain and Ireland, except some island groups. Despite its abundance and bold character, it is not particularly easy to see. It is distinguished from the smaller Weasel by its long, black-tipped tail and a neater dividing line between the red-brown upperside and creamy-white underside. Also, some Stoats moult to a partly or entirely white winter coat ('ermine'). Your best chance of seeing a Stoat in ermine is to head north, in the second half of winter – most northern Scottish Stoats will develop full ermine.

Often, Stoat encounters come as a surprise and are over almost as soon as they begin – not necessarily because the Stoat is afraid, but because it is on a mission. Make squeaky noises to try to catch the attention of this busy little animal.

How to find

Timing Stoats are active year-round. They are most likely to be observed at dawn and dusk, but are also intermittently active through the daytime, especially in summertime.

Habitat Stoats prefer open, low-lying countryside but with some scrub or hedgerows for cover – they are also fond of drystone walls with plenty of gaps and crevices, which are often used as breeding spots. They are probably easiest to find in areas that hold good numbers of their main prey – Rabbits.

Search tips Signs of their presence are rather subtle. The droppings are deposited singly, as with most other carnivores, and are small (about 6cm long) with a twisted shape. The tracks show five slender clawed toes. Spending time in the vicinity of a Rabbit warren should eventually produce some Stoat sightings – if you hear a Rabbit's squeal, there is a good chance a Stoat is making a kill.

WATCHING TIPS

Watch for Stoats along linear features in the landscape – hedges, walls, ridges, field edges, country lanes. Often sightings are brief, but you may be able to attract the Stoat's interest by 'pishing' – making a squeaky kissing noise. Stoats are not particularly afraid of humans and may come very close to investigate, if you keep still. In summer, look out for female Stoats moving their kits and try to monitor where they go. If you locate a family's home patch you could enjoy prolonged views of the kits playing together.

Weasel

Mustela nivalis

Length 24cm

1	2	3	4	5

J F M A M J J A S O N D

This is our smallest carnivore, its size and shape suggesting a mouse that has been stretched out on a rack. It is widespread in Britain, but absent from Ireland, as well as most Scottish islands. Unlike the Stoat it never turns white in winter, and its tail is short and slender with no black tip. Also, its ginger-brown upperside is not neatly separated from its white underside – there are usually some spots of brown encroaching onto the white.

Weasels are hyperactive but often quite unafraid of people. If you happen to be standing in the way of a Weasel intent on a prey trail, it may well run right past your feet rather than looking for another way around.

How to find

◼ **Timing** Weasels may be seen at any time of day and any time of year, but they are most active around dawn and dusk. Daytime sightings are more likely in summer, but they can be easier to spot in winter when there is less vegetation for cover.

◼ **Habitat** All kinds of habitats with good ground cover will support Weasels – meadows, farmland, moor, heath and marsh, even some larger gardens. They prey mainly on small rodents so anywhere that holds good numbers of these is likely to attract hunting Weasels.

◼ **Search tips** Weasels are similar to Stoats in that they often like to follow linear features in the landscape, so watch out for them along walls and hedgerows. Their tracks and droppings are like those of Stoats but smaller – only very soft ground will retain prints as they are so lightweight. If you have land or a large garden, setting trail cams is a good way to determine whether Weasels (and Stoats) are present.

WATCHING TIPS

Like Stoats, Weasels are often seen by chance as they race along a path or wall, soon to disappear into the undergrowth. You may persuade them to pause by making mouse-like squeaks to attract their attention. They are naturally quite fearless, but always in a hurry and hunt mainly in cover. Therefore, having prolonged sightings is really a matter of great patience and good luck – even family groups with young kits are unlikely to stay in full view for long. A Weasel seen in summertime carrying 'prey' is worth a second look, as it may be a female moving a kit. Wait around and you may see her return for the rest of the litter.

Mustela putorius

Polecat

Length 52cm

| 1 | 2 | 3 | **4** | 5 |

J F M A M J J A S O N D

With its pale, dark-masked face, the Polecat is easily identified.

Although Polecats are steadily increasing in the wild, they remain quite localised and very difficult to see – more so than any other mustelid. With good views they are distinctive, stocky, dark mustelids with proportionately short tails, small ears and blunt muzzles. The body fur is dark brown, with a yellow undercoat showing through on the upperside, and the face pale with a black 'bandit mask' around the eyes. However, beware escaped pet ferrets (the domestic form of Polecat) – dark forms are near-identical to wild Polecats. The species' stronghold is in Wales but it is also present in good numbers in the Midlands, the Hampshire/Wiltshire area, and south Scotland. Feral ferrets are established on several Scottish islands.

Super sites

★ Cors Caron NNR, Ceredigion
★ Brechfa Forest, Carmarthenshire
★ RSPB Conwy , Conwy
★ Wye Valley and Forest of Dean, Gloucestershire

How to find

■ **Timing** Polecats are active through the night, especially at dawn and dusk. The likeliest time to encounter one is in summer, when females have to hunt in daylight to keep up with the demands of their growing kits.

■ **Habitat** These mammals use a variety of habitats, from woodland edges to marshland and farmland, and are drawn to Rabbit warrens and farms that hold populations of Brown Rats. They are most likely to be found in low-lying countryside.

■ **Search tips** The best bet is probably to set up before dawn, in an elevated hide that gives good views across suitable habitat. Alternatively, try slow drives along quiet lanes, early and late in the day. The droppings are typically twisted and elongated, and the tracks show five long toes.

WATCHING TIPS

Polecats often have their young in Rabbit burrows, or inside farm buildings. If you are lucky enough to hear of a Polecat den in an accessible site, you should be able to set up a hide and watch them coming and going. The feral ferrets in Shetland and some other Scottish islands tend to be easier to watch than true wild Polecats, but are otherwise very similar in appearance and behaviour. Escapee pet ferrets can often survive very well in the wild, although albino forms are hampered by being both conspicuous and (usually) short-sighted.

Pine Marten

Martes martes

Length 70cm

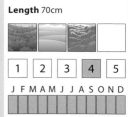

1	2	3	4	5

J F M A M J J A S O N D

Few mammals match the Pine Marten for sheer beauty.

Once very scarce due to persecution, the Pine Marten is increasing and spreading from its stronghold in the Scottish Highlands. It is also now present in north England and Wales, as well as much of Ireland. A large, elegant mustelid, it is easy to recognise by its size, dark fur with large cream bib, large ears and long, fluffy tail. Because they are easily tempted by bait, Pine Martens are quite easy to find and watch in the right areas – a dramatic and welcome change from just a couple of decades ago, when they were much rarer, and regarded as near-impossible to see with any reliability.

Super sites

★ Villages in the Speyside area, Scottish Highlands
★ Speyside Wildlife's private hide on the Rothiemurchus estate, Scottish Highlands
★ Ardnamurchan peninsula, Scottish Highlands

How to find

■ **Timing** Pine Martens are most active at night but can be seen around dawn and dusk, and in summer may be out in full daylight. They are active all year round, and perhaps more likely to come to feeding stations in winter, when natural food is scarce.

■ **Habitat** These animals are accomplished climbers and so prefer habitats with at least some woodland nearby (pine or deciduous) but they will also use moorland, meadows and other more open habitats.

■ **Search tips** By far the easiest way to see Pine Martens is to watch at a feeding station that they are already using. If you are staying in Pine Marten country, try putting out food. Unsalted peanut butter is a favourite and can be smeared onto branches. Eggs and jam sandwiches are also often successful. Walking in suitable woodlands at dusk or dawn may prove successful – remember that the animals are likely to be in the trees. Pine Martens sometimes scavenge at picnic sites, so check these whenever driving in suitable areas at dusk.

WATCHING TIPS

Many guest houses in wildlife-rich parts of the Highlands will have visiting Pine Martens, and there are also paid hides where you are almost guaranteed to see them. Using sticky bait like peanut butter means the martens will stay longer and can be watched at length. Often several will visit the same feeding station at the same time, including babies in summer, and you'll have the chance to watch their charming interactions.

Lutra lutra

Otter

Length 110cm

1	2	3	4	5

J F M A M J J A S O N D

Even out of water, Otters are unmistakeable.

This delightful mammal, once persecuted to near-extinction, is thriving again in our waterways and around our coasts, with the west coast of Scotland holding particularly good numbers. It can be confused with the American Mink, but is much larger and more powerfully built, with a paler coat and a longer, sleeker and tapering tail. It also has a lower profile when swimming, often showing only its head and an arch of tail.

Super sites

- ★ Remote rocky coasts in the Scottish Highlands and Islands
- ★ RSPB Minsmere, Suffolk
- ★ The Norfolk Broads
- ★ RSPB Leighton Moss, Lancashire

How to find

■ **Timing** Inland Otters are largely nocturnal, so most likely to be seen early and late in the day. Coastal animals, though, are frequently active in the daytime as their daily routine is guided by the tides. They are most likely to hunt when the tide is going out, as the falling water levels make their prey more easily accessible.

■ **Habitat** These animals can be seen in rivers, lakes and around the sea coast, and crossing dry land near waterways. They prefer quieter places with less human activity and plenty of prey in the form of fish, crabs and other aquatic life.

■ **Search tips** Scan open water, looking out for a head or whip-like tail. Otters tend to swim at the surface for a few minutes, before diving for several minutes, so be patient. Their presence is revealed by droppings (spraint), which are soft, straight and oily black when fresh, with visible fish remains. Fresh spraint has a sweet but fishy smell, and is often left on prominent rocks near the water's edge. Otter tracks are broad, rather dog-like but five-toed.

WATCHING TIPS

If you spot an Otter swimming at sea, keep still and watch, and use the times when it dives to move closer to the shore. If it catches a large prey item, it will probably come ashore to consume it. Keep still and quiet and you have a good chance of close views. Otters are now regularly seen at several nature reserves and can be watched from permanent hides. Some Otters using rivers that pass through towns can become very relaxed with human presence.

American Mink

Neovison vison

Length 60cm

1	2	3	4	5

J F M A M J J A S O N D

A common North American mammal, this medium-sized mustelid became established in the wild throughout Britain following escapes and deliberate releases from fur farms. It proved a highly damaging and invasive species, particularly problematic for Water Voles and wetland and coastal birdlife, and is regularly trapped and culled. American Minks are usually solid dark brown, often with a little white on the chin and throat, but never with the distinctive face pattern of Polecats. They prefer wetland habitats and may be mistaken for Otters, but are quite different and easily told apart given good views.

The all-dark face helps separate this species from the Polecats. Minks swim strongly and buoyantly, and can climb in trees and up steep river banks – their adaptability allows them to take many different kinds of prey.

How to find

■ **Timing** American Minks are mainly nocturnal, but may be seen in the daytime, particularly in summer and early autumn.

■ **Habitat** All kinds of wetland habitats could support minks. Inland, they prefer well-vegetated lakes, rivers, canals and marshland, while on the coast they prefer rocky shores.

■ **Search tips** Most sightings are brief, chance encounters, much as with the other smaller mustelids. Slowly walking in suitable habitat may pay off, likewise a stake-out in a hide overlooking fresh water – but American Minks are rarely tolerated on nature reserves. Look out for their twisted droppings near water, and their five-toed tracks, which tend to be more splayed out than those of Stoat or Polecat.

WATCHING TIPS

American Minks are much vilified, but in themselves they are as charming and interesting to watch as any mustelid. Young animals can be quite fearless and give prolonged views. However, if you find them living on your local nature reserve it is wise to contact the reserve manager, as minks can be so damaging to native wildlife. There is evidence that Otters will displace minks from riverside habitats, so if Otters are starting to doing well in your local area, you may find that minks become rarer and more elusive.

Meles meles

Badger

Length 90cm

1	2	3	4	5

J F M A M J J A S O N D

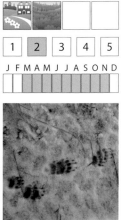

Perhaps our most immediately recognisable mammal, the Badger is also common and widespread, but its nocturnal habits mean many people have never seen one. It is a mustelid, but its stocky grey body and striped face mean it cannot be confused with any other species, and its tracks and signs are also distinctive. Badgers are found throughout Britain and Ireland wherever suitable lowland woodland habitat exists. This animal is highly social – often the whole family will turn up at a feeding station and much charming interaction between them can be observed.

Despite its short-legged physique, a startled Badger can move extremely quickly, so chance encounters are often rather brief. Watching from a hide is the best way to guarantee longer views.

How to find

■ **Timing** Badgers can be seen through spring, summer and autumn, but spend most of their time sleeping in their setts in winter. They are nocturnal but may be encountered at dawn or dusk, in high summer particularly.

■ **Habitat** Setts are usually located in deciduous woodland, in fairly dry areas with soft, sloping ground. The Badgers forage close to the sett but may also wander several miles from the sett at night as they search for food, visiting farmland and gardens.

■ **Search tips** Finding a sett is the first step towards reliably seeing Badgers, unless they are already visiting your garden. A sett comprises several large holes (about 20 x 25cm) in quite close proximity – freshly excavated earth holding broad, five-toed tracks shows that the sett is active. There may also be discarded bundles of old bedding. There will also be latrines nearby – shallow pits in which all the members of the group defecate. Dug-over ground in the garden may be a sign Badgers are visiting.

WATCHING TIPS

If you want to watch a sett, always stay downwind and at a good distance. Arrive before dusk and wait for the Badgers to emerge and leave the area. There are many permanent badger-watching hides in the UK that give great views in full comfort – visit The Badger Trust's website (badger.org.uk) for details. If Badgers visit your garden, you can offer food such as peanuts or dog biscuits to encourage them – scatter the food around and the animals will take longer to find it.

Fox

Vulpes vulpes

Length 110cm

| 1 | 2 | 3 | 4 | 5 |

J F M A M J J A S O N D

With its elegant, leggy build, red coat and long bushy tail, the Fox is both beautiful and unmistakeable. Foxes are very successful animals and can be seen in all kinds of habitats throughout Britain and Ireland, though they are missing from some Scottish islands, and are scarce in some areas because of heavy persecution. Finding and watching these intelligent animals is easy and extremely rewarding. In towns and cities, Foxes can be problematic – raiding bins, menacing outdoor pets and even stealing garden ornaments, but take the right precautions and enjoy living alongside one of our most charismatic animals.

In late summer, young Foxes seek independence and are often easier to watch than the wary, streetwise adults, as they are racing against time to establish territories and find reliable hunting grounds before winter.

How to find

■ **Timing** Foxes are mainly nocturnal but daytime sightings are far from unusual. Dawn and dusk are probably the best times to look. Cubs emerge from earths in early May.

■ **Habitat** Foxes can be found in almost every habitat type, although population density varies quite considerably – sparsely vegetated uplands where prey is scarce hold relatively low numbers. However, many towns and cities support very large populations.

■ **Search tips** Town-dwellers are often made well aware of their local Foxes by the unearthly yowls and shrieks the animals make at night, especially in late winter. Fox droppings are also noticeable by their strong smell. Prey remains are well-chewed, birds' feathers often bitten through rather than pulled out. The breeding earth or den is often well-hidden – it may be a disused hole in a Badger sett, or an enlarged Rabbit burrow. Urban foxes often dig out spaces underneath garden sheds.

WATCHING TIPS

Many householders feed the Foxes that come to their gardens, offering dog biscuits and kitchen scraps. This is a great way to watch Fox activity, but may make you unpopular with your neighbours. Watching Foxes in wilder countryside is highly rewarding. If you have found an active earth, visit (and watch from a distance) at dusk in late spring and you should enjoy views of the young cubs making their first forays into the world. Their playtimes are exhilarating affairs, interrupted periodically by the whole litter dashing back to their increasingly exhausted-looking mother for a quick feed.

Felis silvestris

Wildcat

Length 80cm

1	2	3	4	**5**

J F M A M J J A S O N D

That broad, banded tail is a key Wildcat trait.

Our native wildcat is found only in remote parts of Scotland. It is probably more sought-after by mammal watchers than any other species, but is extremely rare and elusive. The picture is further complicated by the presence of feral domestic cats, and hybrids. Some authorities believe there are probably no entirely 'pure' Scottish Wildcats remaining. A classic Wildcat is a large, thickset cat with dense fur, brown tabby stripes and a broad, blunt-tipped tail marked with neat black rings. Any sighting should be checked and verified with local mammal recorders.

Super sites

★ Ardnamurchan peninsula and other remote parts of mainland west Scotland
★ RSPB Loch Garten, Abernethy
★ RSPB Glenborrodale, Argyle

How to find

■ **Timing** Wildcats may be active at any time of day and night, but are most likely to hunt in open country at night, and to return to the woods by day. They may be easier to spot in winter when vegetation is sparser. Kittens appear any time in late spring or summer, and disperse to new territories in autumn and early winter.

■ **Habitat** Rugged, remote countryside with boulders and areas of woodland is home to these animals. True Wildcats have a strong aversion to human presence – hybrids, though, may be found around villages and farms.

■ **Search tips** The 'best' way to search for Wildcats is to drive slowly through suitable habitat at night, sweeping a spotlight across fields and looking for eyeshine – though any sightings obtained this way are likely to be fleeting and poor. Slow dusk and dawn walks in suitable habitat, pausing often to scan along ridges and other landscape features, may also be worth a try, as may sitting in a parked vehicle overlooking good habitat. Prints and droppings look like those of domestic cats – droppings are often placed in prominent spots.

WATCHING TIPS

Any sighting of a Wildcat is to be treasured. The chances of views lasting longer than a moment are slim. If at all possible, take photographs so that identification can be checked by an expert, but keep location details private to avoid disturbance – pass on details to the local mammal recorder only. Tracking sightings is crucial to help researchers map Wildcat distribution and identify priority areas for conservation. Only with a strong, coordinated effort between landowners, pet owners and conservationists will we stand a chance of saving our only wild cat from extinction.

Grey Seal

Halichoerus gryphus

Length 195cm

1	2	3	4	5

J	F	M	A	M	J	J	A	S	O	N	D

A Roman nose and blotchy spots help identify the handsome Grey Seal.

The larger of the two seals that breeds around Britain, the Grey Seal is an imposing and handsome animal, with a long 'Roman nose' and a usually dark or silver-grey coat with bold black markings. Ninety percent of our breeding Grey Seals are in Scotland, but they can be seen at a few well-known breeding colonies elsewhere, and are also often seen singly all around the coast of Britain, especially in the North and West. At some sites, Grey and Harbour Seals occur together, but can be told apart quite easily when seen side-by-side.

Super sites

- ★ Blakeney Point, Norfolk
- ★ Donna Nook, Lincolnshire
- ★ Farne Islands, Northumberland
- ★ Strumble Head, Pembrokeshire
- ★ Scottish islands

How to find

■ **Timing** Grey Seals may be encountered all year round and at any time of day. They spend much time on land during their spring moult (early spring for males, later for females), and pups are born between September and November – the exact time of pupping varies between colonies. It is vitally important to keep a good distance from the animals at this time, to avoid disturbance.

■ **Habitat** Grey Seals rest on rocky islands and on sandbanks, and they hunt in sheltered seas, often coming quite close inshore.

■ **Search tips** Scan the sea, looking for the protruding head of a seal, often with its snout pointed skywards. Hunting seals tend to loaf on the surface for several minutes, and then make a lengthy dive, surfacing many minutes later and often some distance from their original position.

WATCHING TIPS

Visiting a Grey Seal colony often gives excellent close views as the seals interact on land and in the shallows. When away from breeding grounds the animals are often gentle, curious and tolerant of humans – in some sites such as the Farne Islands it is possible to snorkel or scuba-dive with them, though always allow them to come to you in the water. Watching mothers with pups is a delightful experience but always take care to give them plenty of space.

Phoca vitulina

Harbour Seal

Length 160cm

1	2	3	4	5

J F M A M J J A S O N D

The Harbour Seal has a sweetly expressive, dog-like face.

Despite its alternative name of 'Common Seal', this species is more localised than the Grey Seal and less numerous, with a population of some 55,000, about half that of the Grey. It tends to have a more sandy-toned coat with a finer pattern than Greys, and has a proportionately small head with a more dainty, dog-like muzzle. It often rests with head and hind flippers raised, in a 'banana' pose. Harbour Seals are most numerous in Scotland, particularly the Inner Hebrides, but have outposts at other spots around the British coast. They are less likely to be seen hunting near the shore than Grey Seals.

Super sites

- ★ Blakeney Point, Norfolk
- ★ RSPB Rainham Marshes, Essex (the Thames shore)
- ★ Seal Sands, Teeside
- ★ Skye and other Hebridean islands

How to find

■ **Timing** Harbour Seals may be seen all year round and at any time of day. Their pups are born in early summer and are able to swim within hours of birth. This means that sandbanks that are submerged at high tide can be used as pupping sites.

■ **Habitat** They use rocky shores and sandbanks as haul-out spots, but tend to move into deeper seas than Greys when they are hunting.

■ **Search tips** Visiting well-known haul-out spots is the best way to find Harbour Seals – chance sightings from the shore of hunting animals in the sea are infrequent. The same sites are used repeatedly so once you have found a good spot you can almost guarantee regular observations.

WATCHING TIPS

As with Grey Seals, allow Harbour Seals plenty of space when you visit their haul-out spots, especially during the pupping season. Some of the Scottish haul-outs are easily viewable from land, while others, such as that at Blakeney Point, can be visited by boat. At many sites you can see Harbour and Grey Seals sharing a haul-out, though tending to stay close to others of their own species rather than freely intermingling. Side by side, their differences are quite obvious.

Common and Pygmy Shrew

Sorex araneus, S. minutus

Common Shrew

Length 8cm

| 1 | 2 | 3 | 4 | 5 |

J F M A M J J A S O N D

Pygmy Shrew

Shrews are hyperactive miniature predators, needing to eat more than their own bodyweight in insects and other small prey each day to stay alive. The two common species in Britain are both dull brown in colour and very similar in appearance, easily distinguished from other small mammals by their long narrow snouts that look 'stuck on' to their rounded heads, and their almost invisible eyes. The Common Shrew is 8cm long and is larger-bodied but shorter-tailed. The Pygmy Shrew has a propotionally smaller body and longer tail, although its tail is well over half its body length. Both are common and widespread in Britain, but only the Pygmy Shrew is found in Ireland.

Both of our most frequently seen shrew species are so frantically active that sightings are often too brief to allow identification.

How to find

▪ **Timing** Shrews may be seen year-round and at any time of day or night, but are most active in the warmer months and darker hours.

▪ **Habitat** They are found in all kinds of habitats that have a reasonably thick layer of ground vegetation, and support good numbers of invertebrates. They live in hollows of various kinds, including holes dug out and formerly occupied by other small mammals.

▪ **Search tips** Listen for shrews snuffling and squeaking in the undergrowth. They have very poor eyesight and are not very fearful of humans, but are difficult to spot because they shuttle along their runways very quickly and are often completely concealed by vegetation. Sand traps may reveal their spidery, five-toed tracks. Shrews are protected and may not be trapped without an appropriate licence. If you are carrying out live-trapping for other small mammals, make sure the trap contains food for any accidentally caught shrews – mealworms are suitable.

WATCHING TIPS

These fast-moving, tiny animals are difficult to watch as they prefer to move through cover, and spend much of their time underground. Summer is the time when you are most likely to see interesting activity, such as territorial disputes between rival adults, or families moving in a 'caravan', each holding the base of the tail of the shrew in front, with the mother in the lead. Sometimes, just sitting still in a sheltered grassy area and watching the ground will produce close views of shrews as they scurry along, heedless of everything but their next meal.

Neomys fodiens

Water Shrew

Length 13cm

| 1 | 2 | **3** | 4 | 5 |

J F M A M J J A S O N D

This species is easier to watch than most shrews.

This is our largest shrew species, and the most distinctive, with its velvety black upperside and crisply demarcated white underside. It is also markedly less abundant than Common or Pygmy Shrews, although still very widespread throughout mainland Britain, wherever there is suitable wetland habitat. It is an adept swimmer and finds most of its prey in the water, readily diving to catch small invertebrates such as shrimps, water snails and the larvae of aquatic insects. It also forages on land.

Super sites

- ★ RSPB Fowlmere, Cambridgeshire
- ★ RSPB Rye Meads, Hertfordshire
- ★ RSPB Leighton Moss, Lancashire
- ★ RSPB Blacktoft Sands, East Yorkshire

How to find

■ **Timing** These animals can be seen at any time of day and all through the year, but sightings are most likely in the evenings in late spring and early summer.

■ **Habitat** Mostly found around fresh water with rich marginal vegetation: streams, rivers and lakes. It has a particular affinity with watercress beds. It may also be found well away from water on occasion.

■ **Search tips** Sitting and waiting alongside waterways that are used by Water Shrews is the best way to see one. You may hear their squeaking calls. You can also determine their presence if you manage to find the nest entrance holes (dug into the bank by water but often obscured with vegetation – entrance hole measures about 2cm across) or the piles of crumbly black droppings, containing prey remains, each about 7cm long.

WATCHING TIPS

Water Shrews are fascinating and little-known animals. Like other shrews they have poor eyesight and may well not notice you if you sit still and quiet. If you find a good site with very clear water, you may be able to watch them swimming and hunting underwater. Family groups can be very active and easy to watch in early summertime. Because Water Shrews prey mainly on aquatic insects and other invertebrates, their presence is a sign of clean, healthy waterways that will probably support a good range of other wildlife.

Mole

Talpa europaea

Length 15cm

1	2	3	4	5

J F M A M J J A S O N D

Molehills are easy to find, and their abundance indicates that the animal itself is common and widespread. However, observing healthy Moles in the wild is extremely difficult, as they have little need to surface from their networks of underground tunnels. They are highly adapted to this way of life, with plush dark velvety fur, powerful long-clawed digging paws, and a very sensitive nose. Moles are abundant in Britain, anywhere where the soil is suitable for their diggings, but absent from Ireland.

A velvet coat, pink pig-like nose and those huge spade-shaped forepaws make the Mole quite unmistakable, but you will need a great amount of luck to actually see a living Mole in the wild.

How to find

■ **Timing** Moles are active all year round and potentially at any time of day, but the chances of seeing one at the surface is probably highest in late spring and summer.

■ **Habitat** Molehills are particularly noticeable on otherwise well-kept lawns, but Moles are also present in farmland, meadows and deciduous woodland, and more sparsely in other habitats. They need soil that is not too dry nor prone to flooding, with plenty of earthworms.

■ **Search tips** If you spend some time watching an area with fresh molehills, especially after rain, you may get lucky and glimpse a Mole at work, pushing up soil or pulling nesting material into the soil. After very heavy rain, Moles' tunnels may flood, forcing the animals to the surface. In summer, young Moles disperse from their birthplaces and may make short parts of this journey over land. If you want to have Moles removed from your land, use a firm that traps them alive and relocates them.

WATCHING TIPS

Moles conduct the vast majority of their business below ground, so making any prolonged observations of their behaviour is close to impossible. If there are Moles in your garden and you are willing to allow them to stay, you may over time have some good sightings. Above ground, Moles can scoot along surprisingly quickly as long as their big forefeet can find purchase, and they will quickly bury themselves in leaf-litter if they aren't able to get underground straight away. Captive Moles are sometimes on view at the British Wildlife Centre in Surrey: britishwildlifecentre.co.uk.

Erinaceus europaeus

Hedgehog

Length 22cm

1	2	3	4	5

J F M A M J J A S O N D

This charming, spiny mammal, one of our most beloved species, is present throughout Britain and Ireland. However, it has suffered a significant population decline over the last few decades, due to loss of habitat, loss of prey abundance due to heavy pesticide use, and a high death toll on the roads. In some areas, predation by Badgers may also be a factor. Hedgehogs are more likely to be seen in gardens than in other habitats, and there is a great deal that gardeners can do to help and encourage them, including avoiding pesticides and allowing parts of the garden to grow wild.

This endearing animal is quite unmistakable. It is also a species that can benefit enormously from a wildlife-friendly approach to garden design and maintenance – widespread action of this kind could help stop its precipitous decline.

How to find

■ **Timing** Hedgehogs are nocturnal, and they hibernate through winter – typically from November to late March. Any Hedgehog seen out in the daytime is likely to be unwell and may need assistance.

■ **Habitat** These animals can be found wherever there is good ground cover and plenty of invertebrates, especially worms, slugs and snails. Gardens, woodlands and hedgerows are the best habitats for them.

■ **Search tips** When they are out foraging, Hedgehogs are noisy, shuffling and snuffling along. If you think there may be Hedgehogs in your garden, sit outside on a summer night and listen for this activity. Trail cameras are also useful for locating them. The tracks are five-toed and resemble miniature human handprints. Droppings are about 2–3cm long and full of insect exoskeleton remains.

WATCHING TIPS

If you have garden Hedgehogs, you can encourage them to stay by offering food (cat biscuits are best – avoid bread and milk as this upsets their stomachs) and shelter. Woodpiles and heaps of dry leaves can be suitable for daytime sleep and hibernation, or you can buy purpose-built shelters or 'hogitats'. Hedgehogs are not very fearful of humans but should still be watched from a respectful distance, for the best chance of seeing natural behaviour. Wildlife rescuers will take in Hedgehogs in need of care – this means any found out in full daylight, and any underweight animals (450g or lighter) you may find as winter approaches.

Common Dolphin

Delphinus delphis

Length 2.2m

1	2	3	4	5

J F M A M J J A S O N D

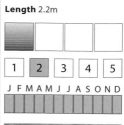

This beautifully marked animal is the most likely dolphin species to be seen in British waters. It is rather a small dolphin with a sleek, elegant silhouette. It has a long, slim dark beak, a dark upperside, and a distinctive hourglass-shaped marking running the length of its flanks – this marking is pale yellow in front of the dorsal fin and pale grey behind it. Common Dolphins are most abundant off the west coast of Britain (especially Scotland) and Ireland, and are seldom seen off east and south-east England.

This small dolphin is one of the most distinctively marked cetaceans likely to be seen in British and Irish seas.

Super sites

★ Try pelagic trips off Cornwall, Wales, Ireland and west Scotland. Check with Seawatch Foundation's list of recommended operators

How to find

■ **Timing** Common Dolphins may be seen all year round and at any time of day, but sightings are most frequent in late summer and autumn.

■ **Habitat** Usually seen well away from land, the Common Dolphin is most likely to be active in calm seas. Gatherings of gannets and other seabirds indicate a lot of fish activity, which may attract dolphins.

■ **Search tips** Taking a boat trip in suitable waters gives a very good chance of encountering pods of Common Dolphins. They are noticeable from a good distance, especially if the sea is calm, as they are very active, and there is a high chance of them approaching moving boats to bow-ride. Summer whale-watching trips off west Scotland offer an excellent chance of close views.

WATCHING TIPS

These dolphins are delightful and exciting to watch, as they frequently breach, jumping fully clear of the water, and bow-ride at exhilarating speed. Their vocalisations may be heard when they come to the surface. Groups encountered in Britain are usually of about 20–30 animals, perhaps including some young calves, but on occasion, much larger pods, of a hundred or more, may be seen. Wildlife photographers on suitable boat trips may want to use a shorter lens than usual, as Common Dolphins are quite likely to come very close.

Tursiops truncatus

Bottle-nosed Dolphin

Length 3.5m

| 1 | 2 | **3** | 4 | 5 |

J F M A M J J A S O N D

An eye-to-eye moment with one of our brainiest mammals.

Familiar to older generations as the star of TV programme *Flipper*, as well as dolphin displays, this large, highly intelligent and energetic dolphin is regularly encountered around some parts of the British coastline and at certain sites can be reliably observed from land. It is mid-grey above, paler on the underside, with the paler colouration extending onto the lower face and forming subtle 'spectacle' markings around the eyes. It has a fairly tall dorsal fin, less hooked in shape than the Common Dolphin's, and a rather short, blunt beak with a protruding lower jaw.

Super sites

- ★ Chanonry Point and other headlands, Moray Firth
- ★ Cardigan Bay, Ceredigion
- ★ Sound of Mull, north-west Scotland

How to find

■ **Timing** Bottle-nosed Dolphins can be seen at any time of year. Some populations habitually shift their feeding grounds through the year – for example, a group present around the extreme west of Cornwall in winter are known to move north in summer, while spring is the best time to see them in the Moray Firth.

■ **Habitat** These dolphins feed closer inshore than Common Dolphins, readily coming into bays and even harbours, though they may be encountered in deep seas as well.

■ **Search tips** Finding Bottle-nosed Dolphins on a whale-watching trip is somewhat a matter of luck, but there are several spots on the mainland where they may be seen regularly from land. Additionally, single animals sometimes linger at particular spots. Following wildlife-watching social media is the best way to keep up with sightings.

WATCHING TIPS

Bottle-nosed Dolphins are less agile than smaller species but still highly entertaining to watch. Those in the Moray Firth can regularly be seen catching prey near the surface, as well as performing spectacular leaps and twists in the air. They readily approach boats to bow-ride. Lone Bottle-noses, usually mature males, sometimes set up home very close to shore. Most famous of them in Britain is Fungie, who has frequented Dingle harbour, Ireland, for (at the time of writing) 32 years.

Watching cetaceans

Britain and Ireland may have relatively few land mammals compared to mainland Europe, but when it comes to marine mammals, we are truly world class with at least 26 species (more than a quarter of the world's total) recorded in our waters. They range from the diminutive Harbour Porpoise to the largest animal that has ever lived – the Blue Whale.

Watching cetaceans from land

In early spring 2017, a Humpback Whale lingered close to the shore off south Devon for several weeks, taking full advantage of unusually high numbers of mackerel and other small fish. A typically extrovert example of her species, she delighted wildlife watchers with frequent energetic breaches and other acrobatics. Encounters like this are infrequent, but a few cetacean species can regularly be seen from land at particular sites – in particular Bottlenose Dolphin, Harbour Porpoise, Orca and Minke Whale.

Any point along the coast could potentially provide cetacean sightings, but headlands will be better than bays, and in general the Atlantic coast is best – particularly north-west Scotland (including island groups). Find a headland giving elevated, well-lit views, and settle down to watch. Sightings are likely to be sporadic so a picnic is a good idea! Binoculars will be adequate for closer sightings, but you may find a

Orcas are oversized dolphins and will sometimes come close to boats to bow-ride.

telescope helpful, especially with an eyepiece that has a wide-angled field of view. Dolphins will be spotted when they surface, but with whales it is likely to be the 'blow' that you notice first. The angle and size of the blow will help you to identify the species, along with what you can see of its back and dorsal fin as it comes up to breathe.

Signs that whales are around include gatherings of seabirds like Manx Shearwaters and Kittiwakes. The birds are attracted by fish disturbed and brought near the surface by feeding whales. Also look out for 'footprints' – briefly lingering circular patches of 'flat' water left by the powerful beating of a whale's flukes as it starts a dive.

Whale-watching boat trips

Many companies offer trips in small boats out into good cetacean spots all through spring, summer and autumn – the crew will be skilled at spotting signs of whale activity. Choose a company with a strong code of conduct, committed to avoiding stress and disturbance to the animals. Seawatch Foundation's recommended code of conduct can be found here: seawatchfoundation.org.uk/ marine-code-of-conduct and they also provide a list of recommended operators: seawatchfoundation.org.uk/ recommended-boat-operators

When on the boat, scan all around you regularly, with binoculars and the naked eye, looking for signs of whale blows and seabird activity – alert the crew to anything interesting that you see. If you are taking a camera, you'll probably find a short zoom lens more helpful than a long telephoto lens, as dolphins in particular may come very close to the boat. Some boats will carry specialist recording equipment so you can see and hear the animals underwater.

Spotting a whale's blow is a thrilling moment for any wildlife watcher.

Strandings

While a few cetacean species habitually hunt close inshore, for many others coming close to land is a bad sign, indicating the animal may be disorientated, injured or unwell. Helping cetaceans in trouble is a job for experts, as it is difficult and potentially dangerous. Rescues are not always successful either – many will remember the young Northern Bottlenose Whale which entered the River Thames in January 2006, showing signs of distress, and died despite rescue efforts. However, the Humpback Whale in Devon, mentioned above, was twice successfully freed from whelk pot lines that had entangled her.

If you find a live stranded cetacean, or one that you are concerned may be in trouble, contact BDMLR (British Divers Marine Life Rescue, bdmlr.org.uk) on 01825 765 546, or the RSPCA on 08705 555 999. The BDMLR will also take reports of dead cetaceans – all reports are helpful for monitoring the state of our whale and dolphin populations.

A **Fin Whale** lies beached on Carlyon Bay Beach near St Austell, Cornwall. The whale had earlier been spotted in St Austell Bay.

This **Sperm Whale** was found struggling in shallow water off Hunstanton, Norfolk, in February 2016.

Atlantic White-sided and White-beaked Dolphins

Lagenorhynchus acutus,
L. albirostris

Length 2.5m

1	2	3	4	5

J F M A M J J A S O N D

Atlantic White-sided Dolphin   White-beaked Dolphin

Risso's Dolphin

The Atlantic White-sided and White-beaked Dolphins are both medium-sized species that may be encountered on boat trips off the north-west coast in particular. Both are dark with white markings on the flanks – White-beaked is the heftier of the two and has a very short, pale beak, while Atlantic White-sided has a stout black beak. Other rarer dolphins that may be encountered in British waters include the Striped Dolphin, medium-sized with a double dark stripe along its pale flanks and a rather long, slim beak, and the large, beakless Risso's Dolphin, with its tall dorsal fin and pale, usually heavily scarred skin.

Risso's Dolphin usually has many scrapes and scratches on its skin.

Super sites

★ Boat trips from mainland north-west Scotland (e.g. Gairloch, Ullapool) and from west Cornwall, especially Bay of Biscay crossings

How to find

■ **Timing** Atlantic White-sided and White-beaked Dolphins are both encountered most often in late summer. Striped Dolphin sightings occur throughout autumn and early winter, while Risso's are more often seen in early summer.

■ **Habitat** All of these dolphin species are most likely to be found well away from land, in deep, cold Atlantic waters, though White-beaked and Risso's Dolphins are also sometimes seen from land.

■ **Search tips** An extended whale-watching boat trip that heads well out into deep seas offers the best chance of encountering these rarer species. You may also be lucky if you watch from a ferry while crossing between north-western islands, while Striped Dolphins may be seen on crossings from south-west England to Spain that cross the Bay of Biscay.

WATCHING TIPS

It is wise to familiarise yourself with the key identification features of the more unusual dolphin species, in case sightings are brief, but the smaller species will often come close to boats to bow-ride (Risso's rarely does this, though). Striped Dolphins are particularly active and entertaining to watch. Photographing dolphins may be easy or very difficult, depending how lucky you are on the day. A skilled skipper will be able to get you closer under most circumstances, without causing any stress to the dolphins.

Ziphius cavirostris,
Hyperoodon ampullatus

Cuvier's Beaked and Northern Bottlenose Whales

Length 6–8.5m

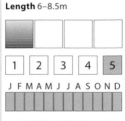

| 1 | 2 | 3 | 4 | 5 |

J F M A M J J A S O N D

Cuvier's Beaked Whale

Northern Bottlenose Whale

The smaller whales recorded in British waters include several species of beaked whales, with Cuvier's Beaked Whale by far the most frequent, as well as Northern Bottlenose and Long-finned Pilot Whales. There have also been a handful of sightings of other smaller cetaceans including Narwhals, False Killer Whales, Pygmy Sperm Whales and Belugas – these are probably best regarded as vagrants to our region. Seeing any of these species is difficult to say the least, although whale-watching is nothing if not unpredictable. Identification can be difficult – take photos, if possible, of back profile and blow shape.

These small whales are not easy to find but can give excellent views if you are lucky enough to encounter them on a pelagic trip.

Super sites

★ Boat trips from west Cornwall, especially Bay of Biscay crossings, and mainland north-west Scotland

How to find

■ **Timing** Unusual whales may be encountered at any time of year, but for most species the summer months are more likely. However, trying to spot marine mammals in the Bay of Biscay can be productive in winter, and Long-finned Pilot Whales are more often seen in winter than summer, especially further south.

■ **Habitat** These species favour deep water, and any seen close to land are likely to be experiencing problems. There are exceptions though – in 1996 a Beluga took up residence in a sea-loch in Shetland, giving excellent views to watchers on land.

■ **Search tips** Long ferry crossings probably offer more chance of seeing these species than the typical 2–3-hour whale-watching trip, though the advantages of the latter are that the boat will stop if a whale is seen, and that there will be experts on hand. Experience the best of both worlds with a two-night whale-watching cruise from Portsmouth to Santander and back to Plymouth with Brittany Ferries – visit brittany-ferries.co.uk/offers/mini-cruises for details.

WATCHING TIPS

Although large and hefty, these whales can be energetic and dramatic to watch, with Long-finned Pilot Whales particularly likely to breach, and also particularly sociable, even swimming with dolphins at times. These are true deep-sea animals and, with so much ocean for them to range around, you do need good luck to encounter them. If you have any unusual sightings, you can notify Seawatch Foundation online: seawatchfoundation.org.uk/sightingsform.

Harbour Porpoise

Phocoena phocoena

Length 1.5m

| 1 | 2 | 3 | 4 | 5 |

J F M A M J J A S O N D

One of the smallest cetaceans in the world, this delightful little porpoise can be observed fairly easily around the British coast, and often close inshore, but is easily overlooked because of its unobtrusive behaviour. Views tend to be limited to a hump of grey back, showing a small, broad-based dorsal fin. The head and tail rarely break clear of the surface, but if you do see its head you'll notice a rather blunt and featureless face, with small eyes and no beak or forehead bulge. At close range the blow has a quiet puffing sound. Harbour Porpoises are quite common around Scotland, Wales, Ireland and most of England, but scarcer in the east and south-east.

This endearing, blunt-nosed little cetacean rarely shows itself this clearly, but its willingness to venture very close to shore does make it probably our easiest cetacean to watch from land, in most areas at least.

How to find

■ **Timing** Spring and late summer–autumn are the best seasons, but Harbour Porpoises may be observed at any time of year.

■ **Habitat** They prefer shallower water so are often seen close to the shore, and may take refuge in sheltered bays when conditions are stormy at sea. Fish concentrations that attract flocks of seabirds may also attract some porpoises.

■ **Search tips** It is difficult to pick out the small grey back of a porpoise rising among waves on a choppy sea – so calm days are best to seek out this species. If watching from land, find a somewhat elevated viewpoint. Harbour Porpoises are regularly seen on short ferry crossings, which may give you the opportunity to see them from above and have a better or more complete view than you would from shore.

WATCHING TIPS

Harbour Porpoises may lack the acrobatic energy of their dolphin cousins, but are still replete with charm and delightful to watch as they dive and surface in their gently rolling action. You will often encounter them in small groups of up to eight animals – with young calves in the summer months. Any time you make a boat crossing, anywhere around Britain but particularly when 'island-hopping' in Scotland, look out for porpoises – especially close to port.

Orcinus orca

Orca

Length 6–9m

1	2	3	4	5

J F M A M J J A S O N D

There is no mistaking that piebald head for any other cetacean. Orcas often show themselves very clearly and are curious, highly intelligent and opportunistic predators.

Unmistakeable, dramatic and still somewhat feared, the Orca or Killer Whale is actually a member of the dolphin family, and like other dolphins is an energetic and intelligent hunter of large prey. The tall and (in adult males) very straight dorsal fin makes it easy to identify. If the black-and-white body pattern is not seen, you are likely to get a clear view of the white eye patch as the whale comes up to breathe. It is social, usually seen in small pods, and its very versatile hunting behaviour means that it may come much closer inshore than other large cetaceans do. It is most likely to be seen off north and north-west Scotland and west Ireland, with the Shetlands the best region in Britain.

Super sites

★ Shetland

How to find

■ **Timing** It may be seen at any time of year, but there are relatively more sightings between June and October than at other times.

■ **Habitat** Orcas may be found in deep seas but also close to land, in sea lochs and bays. They are attracted to seal breeding colonies.

■ **Search tips** You may have a chance encounter with an Orca from almost any whale-watching boat or ferry crossing, but chances improve dramatically in the Shetlands. From land, look out for them loitering in the vicinity of seal colonies, and keep an eye on wildlife-watching social media for news of groups lingering close to the shore. Orcas sometimes approach and swim alongside boats.

WATCHING TIPS

Its sheer size and power makes the Orca breathtaking to watch, and that added to its dolphin-like exuberance can make for some thrilling encounters. In Shetland, you may witness a pod of Orcas hunting seals, using a powerful blow from the flukes to toss a hapless victim into the air before seizing it. Other behaviours you could witness include breaching, tail-slapping and spy-hopping (tilting the head out of the water to examine its surroundings above the surface).

Minke Whale

Balaenoptera acutorostrata

Length 7.5m

1	2	3	4	5

J F M A M J J A S O N D

This is the smallest and most common of the baleen or great whales that you are likely to find in British waters. With good views you will see the long, pointed, triangular head, with throat pleats. On surfacing, the Minke Whale shows a long dark grey back with a small, hooked dorsal fin about two-thirds of the way along – its tail never comes clear of the water when it dives. Its blow, given just a moment before the fin surfaces, is vertical and can reach 3 metres high but is still rather difficult to spot. Minke Whales are most likely to be seen off Scotland and west Ireland, and may be spotted from shore in certain areas.

Report stranded whales (dead or alive) on the CSIP (Cetacean Strandings Investigation Programme)'s stranded whale or dolphin hotline: 0800 652 0333.

Super sites

★ From headlands and boat crossings in north-west Scotland and Hebrides

How to find

■ **Timing** Most sightings are between May and October, with many animals feeding particularly actively in autumn as they prepare to migrate. However, Minke Whales may be encountered at any time of year.
■ **Habitat** These whales are usually seen well out at sea, but sometimes come close enough inshore that they can be seen from land, especially from headlands over deep water.
■ **Search tips** Flocks of shearwaters or other seabirds may reveal the presence of a feeding Minke Whale. Scan the sea with binoculars or a telescope, looking out for the blow and surfacing back of the whale. On land, position yourself with the sun behind you for the clearest views, and scan regularly at all distances.

WATCHING TIPS

Minke Whales offer the best chance in Britain to observe a 'great whale' in action, and can be energetic feeders, sometimes jumping partially clear of the water. They may approach boats, giving close views from above. Often they are seen in the midst of flocks of feeding seabirds, offering a grand spectacle of marine wildlife in the same view. If you spot a mass of Manx Shearwaters feeding out at sea, keep scanning the area for signs of whale activity – if the light is good and the sea not too choppy it should be possible to spot the blow and rising back of a feeding Minke.

Balaenoptera physalus,
Megaptera novaeangliae

Fin and Humpback Whales

Length Fin (19m)
Humpback (13m)

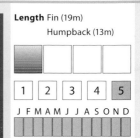

1	2	3	4	5

J F M A M J J A S O N D

Humpback Whale

Fin Whale

The Fin Whale is the largest cetacean regularly seen in British waters. It is long-bodied with a small, backwards-angled dorsal fin, and a white area on the lower part of the face. The Humpback is very distinctive with its very long flippers, lumpy skin texture, and energetic breaches and rolls. Other rarer species include the Sperm Whale, a predator with a distinctive massive bus-shaped head; the Sei Whale, a slightly smaller close relative of the Fin Whale; the North Atlantic Right Whale, a very stocky animal with a large head and no dorsal fin; and the Blue Whale, a slim whale that may approach 30 metres in length. None are easy to see, although Bay of Biscay crossings offer a fair chance of Fin Whale sightings.

Usually the only surfacing part of a whale you'll see is the back, but 'spy-hopping' (raising the head above water) is not uncommon.

Super sites

★ Any sea crossings towards and through the Bay of Biscay

How to find

■ **Timing** Fin Whales are more often seen between April and December, while Humpback sightings happen at all times of year. Sei Whales have recently been observed annually off west Ireland in autumn.

■ **Habitat** These are animals of deep seas, although Humpbacks may feed closer to shore. Any whale lingering very close to shore should be watched for any signs of distress.

■ **Search tips** Long, late summer ferry crossings in the south-west and north-west offer the best chance of encountering any of these rare whales. Spend as much time as you can on deck, watching out for seabird activity and scanning the waves for blows.

WATCHING TIPS

The occasional Humpback Whale found in British waters is likely to offer exciting viewing opportunities, as this species is particularly exhibitionist, regularly breaching and showing off its distinctive long flippers. The other great whales are much more sedate. They will tend to surface maybe half a dozen times in quick succession, in between dives that last at least 10 minutes, so this is the time to study their movement and appearance and pinpoint their identification.

Pipistrelles

Pipistrellus

Common Pipistrelle

Length 5.5cm

J F M A M J J A S O N D

Nathusius's Pipistrelle

In the 1990s, the pipistrelle bats in Britain were found to be two species rather than one, after analysis of their call frequencies. The Common and Soprano Pipistrelle are both very common species throughout Britain and Ireland, the Soprano dominating in the north and the Common further south. Nathusius's Pipistrelle, slightly larger and fluffier than the other two, is a rare species – formerly only known as a migratory visitor but now known to have a few breeding colonies in Britain and Ireland. All three pipistrelles are small bats that often occur in urban environments, though you will need a bat detector to identify them to species level.

The only scarce pipistrelle in Britain is Nathusius's.

Super sites

Nathusius's Pipistrelle
- ★ Bedfont Lakes CP, London
- ★ Stockgrove CP, Beds./Bucks.
- ★ Clotworthy Arts Centre, Antrim, Co. Antrim
- ★ Ardress House (National Trust), Co. Armagh

How to find

■ **Timing** Pipistrelles are active from mid-spring until mid-autumn, depending on temperature. They emerge from their roosts relatively early in the evening.

■ **Habitat** Roosts are often large, and usually inside buildings, also sometimes hollow trees or caves. Bats found inside homes are more likely to be either Common or Soprano Pipistrelles than other species. The bats hunt over gardens, woodland and, Soprano and Nathusius's in particular, often over open water.

■ **Search tips** A pleasant way to watch for pipistrelles is to sit outside on a warm summer evening as the sun goes down. Look out for very small, fast-flittering bats hawking for insects around trees and in the open air. If using a bat detector, Common Pipistrelles' calls register most clearly at 45kHz, Sopranos at 55kHz, and Nathusius's at 38kHz.

WATCHING TIPS

Common and Soprano Pipistrelles are the easiest species to spot in towns and villages. Look out for the first youngsters of the year in mid-summer. In late summer, males fly up and down a linear path when seeking a mate, giving loud calls. Hunting activity reaches its peak in early autumn as the bats prepare to hibernate. Maintaining a natural garden will help ensure a good supply of flying insects for them, and you could also consider putting up bat boxes on your walls or garden trees. Nathusius's Pipistrelle is only known from a few sites so far but may be under-recorded.

Nyctalus noctula

Noctule

Length 8cm

| 1 | 2 | 3 | 4 | 5 |

J F M A M J J A S O N D

This is our largest bat (though still a tiny animal) and has relatively long, narrow wings and a powerful flight. Its echolocation and social calls are often audible to the (younger) human ear. Noctules are relatively common and widespread in England, Wales and southern Scotland, but are absent from Ireland. Their distinctive strong, fast and swooping flight and long-winged outline, tendency to emerge early in the evening and preference for hunting in open habitats make them one of the easier species to find and identify.

The Noctule is a formidable predator, as bats go. Its relative, the Greater Noctule, is Europe's largest bat and regularly catches birds in flight – luckily for British birds, it has not been recorded here.

How to find

■ **Timing** Noctules are on the wing through spring to autumn, and may emerge on mild winter days. They leave their roosts early in the evening, and may be out in full daylight in mid-summer.

■ **Habitat** Summer roosts are mainly in tree holes – they will also use bird nestboxes and bat boxes, and rock crevices. They hibernate in similar places – winter gatherings are normally small. They hunt in all kinds of open habitats, often flying very high.

■ **Search tips** Look for Noctules zooming around trees and field edges from early evening, and listen too (if you are older and have trouble picking up high-pitched sounds, persuade a younger friend to help you!) Echolocation calls are loudest at 25kHz.

WATCHING TIPS

These bats are exhilarating to watch as they race and dive in pursuit of large flying insects. Although less common in urban environments than the pipistrelles, Noctules may be attracted to gardens that support plenty of moths, beetles and other good-sized insect prey – they may also use tree-mounted bat boxes. Because they will fly in good light, with binoculars you should be able to see them in detail, and, perhaps in July, see females carrying their small babies in flight.

Bats

Britain has more species of bats than any other mammal group, and these 18 species are very popular, if challenging, quarry for the keen wildlife watcher. Some are common, others very scarce, but all are strictly protected from disturbance. It is therefore very important to approach bat-watching with care.

Bat roosts

Bats may roost singly or in groups – sometimes large groups. For example, in 2015 a roost of some 2,500 Soprano Pipistrelles and Brown Long-eared Bats was found inside a rural cottage roof-space in Norfolk. All bat roosts are strictly protected from disturbance at all times, which means you cannot legally enter a roost, unless accompanied by a bat licence holder. Nor can you use a torch to light up the inside or entrance to a roost. However, it is possible to see the animals leaving the roost at dusk if you watch from a reasonable distance. There are also a few roosts around the country which you can watch via webcam – for example Greater Horseshoe Bats in Devon at devonbatproject.org, and Lesser Horseshoe Bats in Denbighshire at chesterzoo. org/explore-the-zoo/attractions-and-exhibits/web-cams/bat-cam.

If bats are using your garden or property as a roosting site, you have a great opportunity for good views, although you are still legally obliged to avoid disturbance at all costs. Try putting up bat boxes to

Bat boxes are open at the bottom, allowing the bats to climb up into them.

encourage more visitors – place them on different sides of the same tree trunk, as the temperature inside each box will be different depending on how much sunlight exposure it has, and the bats can select the one that suits them best at different times of year. Gardens that support plenty of insects will attract hunting bats, so keep things a little wild and allow nature to thrive.

If bats' presence is causing problems, contact the Bat Conservation Trust (bats.org.uk), who will put you in touch with a local bat group for advice. They can also offer advice and help if you find an injured bat, or if you discover a bat trapped in your home.

Pipistrelles are tiny bats and can crawl into very narrow spaces to roost.

With its rams' horn ears, the **Brown Long-eared Bat** is identifiable at a glance.

Greater Horseshoe Bats roost in closely packed groups.

Using bat detectors

All of our bats hunt and navigate using echolocation. Their calls are, almost without exception, beyond our range of hearing, but through the use of bat detectors we can not only pick up these ultrasound vocalisations but often identify the calling bat to species level. There are various kinds of bat detectors – all work by detecting the ultrasound calls of bats and converting them to frequencies audible to human ears. Each species has a 'peak frequency' – the frequency at which its calls are loudest, and the form of calls is also often distinct – for example, the 'tick-tock' of the Noctule, or the 'warbles' of horseshoe bats. Some bat detectors make recordings, and can be connected to computer software to convert sounds into sonograms – visual representations of the sound, showing changes in frequency plotted against time. These can be very helpful with identification.

However, it takes practice to use any bat detector to interpret the calls you pick up (and to distinguish them from other ambient noise), and even the most basic models are quite expensive. If you take part in some of the many bat walks that take place on nature reserves throughout the summer, you'll see them in use in expert hands, and you'll also have the opportunity to try before you buy.

Using bat detectors will reveal bats that you may not be able to see, and perhaps give you an insight into the sensory world of bats – shaped by sound rather than anything visual. Actually seeing bats in full darkness is

A family out in the field using **bat detectors** to listen for the high-pitched clicks of the elusive animals as they hunt in the night sky.

very difficult – they fly too quickly to be picked up in torchlight, and in any case, shining a torch at them could cause disturbance. If watching in the evening or on moonlit nights, you may have some success with binoculars. Choose models that have large objective lenses for the best low-light performance.

Greater and Lesser Horseshoe Bats

Rhinolophus ferrumequinum,
R. hipposideros

Length Greater (10.5cm)
Lesser (7cm)

1	2	3	4	5

J F M A M J J A S O N D

Greater Horseshoe Bat

Lesser Horseshoe Bat

Our two species of horseshoe bats are close relatives, sharing a remarkable face structure with a prominent fleshy 'nose-leaf', which helps them to generate their complex echolocation calls. Both species are uncommon and restricted to south-west England and Wales (and south-west Ireland in the case of the Lesser Horseshoe Bat). They mostly roost inside buildings or caves. On a bat detector, the Greater Horseshoe's calls are loudest at 80–85kHz compared to the Lesser Horseshoe's 110kHz, and both species' vocalisations are interpreted as a continuous, musical warbling, quite unlike the sounds made by any other British bats.

Super sites

Both
★ Around Dartmouth, Devon
Greater
★ Buckfastleigh, Devon
★ Minster Church, Cornwall
★ Stackpole Estate, Pembrokeshire
Lesser
★ Arlington Court, Devon
★ Coed y Brenin, Gwynedd

How to find

■ **Timing** These two species do not leave their roosts until late in the evening: at least half an hour after sunset. They are active from April or May to September.

■ **Habitat** Roosts are most often in old, undisturbed buildings, but also sometimes in caves. These bats prefer to hunt in woodland, open meadowland and along hedgerows and other linear landscape features, sometimes flying quite close to the ground.

■ **Search tips** Using a bat detector is the easiest way to locate these species. Greater Horseshoe Bats are vocal at their roosts in summer, and may also leave accumulations of prey remains under favourite feeding perches. Your local bat group should be able to advise you about sites, and it may be possible to join licensed bat workers on site surveys or roost visits.

WATCHING TIPS

Watching horseshoe bats is difficult as they are both late-emerging species, but if you visit a known site on a brightly moonlit night you could be lucky. Greater Horseshoe Bats have interesting feeding behaviour, often sallying out from a perch to take prey and returning to the same spot to land and consume their catch. Lesser Horseshoes sometimes glean prey from tree foliage. Both species may hunt at very low levels. Joining your local bat group and working towards obtaining a handling licence may in due course afford you the opportunity to see these remarkable animals close-up.

Plecotus auritus

Brown Long-eared Bat

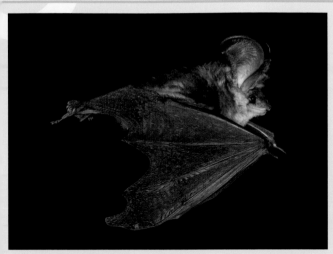

Length 9cm

1	2	3	4	5

J F M A M J J A S O N D

The Brown Long-eared Bat is one of the so-called 'whispering bats', because its calls are so soft. It often picks insects from tree foliage, finding them by listening for the noises they make.

This is perhaps our most distinctive bat species, with its absurdly large ears (about three-quarters as long as its body). It is also relatively common and frequently observed – it is one of the likeliest species to be found roosting in the loft space of a house, for example. At roost, the ears may be curled back or even tucked under the wings (leaving the pointed inner part of the ear, the tragus, sticking up in view), but they are upright in flight and noticeable given a good view. The broad wings and relatively slow but agile flight are also eye-catching. Brown Long-eared Bats occur throughout Britain and Ireland, though are sparsely distributed in the Highlands and are absent from most northern Scottish islands.

How to find

■ **Timing** Brown Long-eared Bats leave their roosts late in the day. They are active from mid-spring to mid-autumn, with most young born in early July.

■ **Habitat** In summer, these bats mostly roost inside buildings, in small colonies, but also in trees. They prefer to hibernate in cold caves and tunnels.

■ **Search tips** The sounds this species makes on a bat detector are very quiet, fast clicks, which are loudest at 40–45kHz but cannot reliably be told from some other species. Their presence may be detected by accumulations of moth wings and other large prey remains underneath favourite perches.

WATCHING TIPS

This bat is a versatile feeder. If you are lucky and the night is bright, you could see it not only chasing prey in open air, but pouncing on beetles on the ground, or taking insects from tree leaves and branches. It is so agile that it can even catch moths flying against lit windows without coming to any harm. It may visit your garden if food supplies are good and there are suitable roost sites nearby – it will use bat boxes. Roosting Long-eared Bats in buildings are often overlooked because their habits are very quiet and discreet.

Natterer's and Daubenton's Bats

*Myotis nattereri,
M. daubentonii*

Serotine Bat

Length 5cm (both species)

1	2	3	4	5

J F M A M J J A S O N D

Natterer's Bat

The remaining bat species found in Britain and Ireland include common species such as Natterer's, Daubenton's and Whiskered, through to real rarities such as Barbastelle and Bechstein's Bats. We are continuing to learn more about our bats every year – it was only in 2010 that Alcathoe's Bat was recognised as a distinct species. Some bats are identifiable to species if you pick them up on a bat detector but others are not. You can find descriptions and information on distribution on the Bat Conservation Trust's website bats.org.uk and on Bat Conservation Ireland's website batconservationireland.org.

Bats are very diverse, and are intelligent, long-lived and socially complex animals. Their habits make them very difficult for the layperson to study, but it is well worth joining your local bat group for the chance to see and learn more about them.

How to find

■ **Timing** Most of these bats emerge from their roosts after sunset. Those most likely to appear in late summer daylight are Serotine, Leisler's Bat and Barbastelle. All are active from spring to autumn, with activity highest in late summer and early autumn as they feed-up prior to hibernation.

■ **Habitat** Some bats have particular hunting and roosting habitat preferences. Daubenton's is known for hunting mostly low over lakes and rivers and even picking prey from the water's surface, while Serotine often hunts over pasture and will catch prey on the ground. Barbastelle and Bechstein's Bats are both very unlikely to be found roosting in buildings, preferring trees within woodland.

■ **Search tips** Heading out on a summer's night, armed with a bat detector and a torch to light your way is an exciting wildlife-watching adventure. Also look out for feeding signs, but remember never to disturb an active roost. It is also very pleasant to just relax in your garden with the detector as the sun sets, and see what you can pick up overhead.

WATCHING TIPS

Joining a bat walk with experts in charge is a great way to watch bats and learn more about them. Search local nature reserves and bat groups to find events near you. The Bedfordshire, Cambridgeshire and Northants Wildlife Trust even offers a bat tour on the Cam by punt wildlifebcn.org/events/bat-punt-safaris. Joining your local bat group will provide wonderful opportunities to learn more and see bats close-up, as well as help with conservation projects to safeguard these amazing mammals.

Vipera berus

Adder

Length 70cm

1	2	3	4	5

J F M A M J J A S O N D

This beautifully marked snake is widespread in Britain (not Ireland), but it is declining. It is also our only venomous snake, but is not aggressive – it will only bite as a last resort if threatened and cornered. Males are light silvery-green, females red-brown (and larger) – both sexes have a bold dark or black zigzag pattern running the length of the back. There is a broad dark line on the side of the face, either side of the red, slit-pupil eye. Melanistic forms are occasionally observed. Adders are protected by law – it is illegal to kill or harm them.

Melanistic Adders are almost pure black, and very striking indeed. The typical male form is our most strongly patterned snake, but its black zigzag pattern on an almost whitish background provides surprisingly good camouflage.

How to find

■ **Timing** Male Adders emerge from hibernation in March, or even February; the females up to a month later. On warm days, Adders can be seen basking in the open, but when the temperature is too high they will retreat to shady cover to keep cool.

■ **Habitat** This is a species most associated with dry, open countryside, such as heaths and moors, but will also occur in woodland and in fields on woodland edges. It is rarely found in gardens.

■ **Search tips** Look for Adders early in the day and early in the year, walking slowly and carefully in suitable countryside (they are very sensitive to ground vibration). Check open sheltered spots close to cover. On some nature reserves, refugia such as sheets of corrugated metal are placed as shelters for reptiles and Adders may use these. It may be possible to join a reserve warden on a survey walk.

WATCHING TIPS

Adders are fascinating snakes with complex courtship and territorial behaviour. Your best bet for observing interesting interactions is to pick a spot at a good site on a warm day in spring, and wait. Males establish territories soon after emergence, and battle with each other in a graceful 'dance', moving together at high speed and trying to push out their rival. Once females emerge in mid-spring, males attempt to guard them from rivals until they are ready to mate. Eight or so live young are born in late summer. Always give Adders plenty of space, for their wellbeing and your own. If you (or your dog) is bitten, seek help immediately.

Smooth Snake

Coronella austriaca

Length 65cm

1	2	3	**4**	5

J F M A M J J A S O N D

A side-on view of the head reveals the distinctive dark eyestripe.

Our rarest snake, this species is found in south-east Dorset and south-west Hampshire, and a smaller area of east Hampshire and west Surrey – it has also recently been reintroduced to sites in east Devon and West Sussex. It is slim and lacks bold, obvious markings, having a double row of small dark spots on a green (female) or brown (male) ground colour. The head is a shade darker, and has a narrow dark stripe running through the pale, round-pupil eye. Its scales are small and lack ridges, giving a smoother, more polished appearance than the Adder or Grass Snake. Smooth Snakes are strictly protected from disturbance of any kind – a fact which wildlife watchers must keep in mind when searching for them.

Super sites

- ★ RSPB Arne, Dorset
- ★ Studland Heath, Dorset
- ★ Avon Heath CP, Dorset

How to find

■ **Timing** Smooth Snakes emerge in early spring, and quickly seek partners and mate. Later in the year they spend most of their time hidden in cover.

■ **Habitat** This is a species of warm, open heathlands, most often dry, south-facing slopes with tall heather; it is also found in patches of tussocky moor-grass (*Molinia*).

■ **Search tips** Your best chance of seeing Smooth Snakes in the open is to visit a suitable site on a warm, still morning in early or mid-spring, or join a reptile walk on a nature reserve. This is a vulnerable and protected species and occurs in fragile habitats, so avoid wandering off paths when seeking it out.

WATCHING TIPS

Smooth Snakes are difficult to see and to watch, and it is vital that you do not actively search them out because of their protected status and high sensitivity to disturbance. Pick a suitable day, find a good vantage point and wait and watch. With good luck you may encounter pairs coming together in spring and even mating. The species is not very active once mating has been accomplished. It preys on other reptiles, which it tends to catch out of sight, deep in the heather – it feeds infrequently. Reptile walks at nature reserves and country parks, led by licensed experts, probably offer the best chance of sightings.

Natrix helvetica

Barred Grass Snake

Length 115cm

1	2	3	4	5

J F M A M J J A S O N D

A frightened Barred Grass Snake may adopt a dramatic, open-mouthed 'playing dead' posture to discourage you from approaching it. Don't be tempted to pick it up – its bite may not be venomous but is very painful.

Our largest snake, the Barred Grass Snake was formerly known just as the Grass Snake, but in 2017 was found to be a separate species to the (Common) Grass Snake found in eastern Europe. It is a slim snake with a dark grey-green upperside, marked with bold black vertical bars on its flanks, and an off-white belly. It has a prominent yellow, dark-edged collar, and a dark eye with a round pupil. This snake is widespread in England, Wales and southern Scotland, but scarce further north and absent from Ireland. It has declined in recent decades but is still relatively common; along with the Slow Worm, it is probably the likeliest reptile to be found in gardens.

How to find

■ **Timing** These snakes emerge from hibernation in early spring but stay near their hibernacula - sheltered places in which reptiles or other animals hibernate through winter - for a few days before beginning to roam further afield in search of prey and a mate, and they may then cover considerable distances. Eggs are laid in June or July and hatch about 10 weeks later.

■ **Habitat** This snake is most often found near lakes, ponds, ditches and rivers, in damp long grass or marshes. It is often seen swimming across still or slow-flowing open water.

■ **Search tips** Look out for Barred Grass Snakes lying in sunshine in early spring, sometimes in small groups. Throughout spring and summer, scan open water for swimming snakes – the head and neck is held up clear of the water. You might find the snakes, and perhaps their eggs, in your compost heap – if so, leave them undisturbed.

WATCHING TIPS

Spending time near water at sites with plenty of amphibians may result in rewarding views of Barred Grass Snakes hunting their prey. Fish are also taken early in the year. If you have a garden pond, or live near areas of open water, you may attract these snakes to your garden – make sure there is plenty of long grass, and some shady hiding places for them. Compost heaps attract snakes seeking warmth and females ready to lay their eggs. Avoid attempting to handle them – when alarmed they will bite and eject very bad-smelling droppings.

Common Lizard

Zootoca vivipara

Length 16cm

1	2	3	4	5

J F M A M J J A S O N D

This is the only 'legged lizard' to be found in most of Britain, and the only native reptile you'll see in Ireland. A relatively small lizard, it is rather inconspicuous, but still quite easy to see because of its fondness for basking in sunshine – it is also usually quite tolerant of approaching humans, as long as you are careful. Common Lizards are variable but usually rather drab in colour, with subtle darker markings. They are sometimes confused with newts, when the latter are found away from water – but the obvious scales on the body and long slender toes will immediately tell you that you are looking at a lizard rather than a newt.

When you are walking along a warm dry path, step softly and keep checking the path ahead for basking lizards. It's not unusual to encounter several basking close together in the same area.

How to find

■ **Timing** Male Common Lizards emerge from hibernation in March, and in April and May they mate with the females, which emerge a few weeks later. About seven live young are born in July. Common Lizards are active mainly in daylight – you're most likely to see them in the open on sunny days.

■ **Habitat** They can be found in a wide range of habitats, including fields, moorland, farmland, sometimes gardens, wasteland and woodland edges – anywhere sheltered but sunny, with plenty of insect prey.

■ **Search tips** Move slowly and quietly in search of Common Lizards. They will bask in the open on warm (less so on baking hot) days, often choosing a log, rock or wooden bench (they are good climbers), but almost always the preferred basking spot is a short dash away from cover. A warmed-up Common Lizard can run very fast – a sudden rustling at a path edge as you walk along may well be a lizard rushing for cover. Try waiting quietly, and it may come out again.

WATCHING TIPS

Common Lizards are easy to see when they are basking, and as long as you move slowly and carefully you can have very close views. Watching them in 'active' mode is more difficult – much of their hunting takes place within the undergrowth, out of sight. Spend time waiting and watching quietly in habitats with plenty of small areas of open or bare sun-warmed ground for the best chance of seeing them dashing after prey.

Lacerta agilis

Sand Lizard

Length 18–20cm

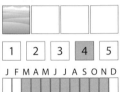

1	2	3	4	5

J F M A M J J A S O N D

A breeding-condition male Sand Lizard is splendidly colourful.

This lizard is rare and localised, occurring on heathland in Dorset, Hampshire and Surrey, and sand dunes in parts of Merseyside and Lancashire. It has also been successfully reintroduced to sites in north Wales, Berkshire, Cornwall and Devon. It is larger and more robust-looking than the Common Lizard, with a larger head. It is also paler, with dark speckling or bold dark spots with pale centres. Its basic colours are very variable, with shades of sandy-brown, grey and green; males in breeding condition can be bright vivid green. Like our other rare native reptile, the Smooth Snake, the Sand Lizard is strictly protected from disturbance at all times.

Super sites

★ RSPB Arne
★ RSPB Farnham Heath
★ Studland Heath, Dorset
★ Freshfield Dune Heath, Merseyside
★ Talacre Dunes, Flintshire

How to find

■ **Timing** Sand Lizards are diurnal, and not typically active until late March or early April. They return to hibernation in mid-autumn. Males are at their most vividly colourful in mid-spring. Females lay eggs in early summer; these are buried in sand and the young hatch in late summer.

■ **Habitat** These lizards are found on warm, sunny heathlands and coastal sand dunes.

■ **Search tips** Pick a warm, not too breezy day, and walk slowly through suitable habitat, pausing to check sunlit spots for basking lizards. As always with searching for reptiles, step softly to avoid alarming the animals with the vibrations of your footsteps. These lizards are not as fast-moving nor as agile as Common Lizards, so can be seen well if you are very careful. Guided reptile walks at nature reserves will offer a good chance of sightings.

WATCHING TIPS

The most interesting behaviour can be seen early in the season. Soon after emergence, male Sand Lizards establish territories, and battles between rival males may be seen at this point, with the combatants wrestling and attempting to bite each other's heads. A female entering a male's territory will be chased and grabbed by the tail or body before mating occurs. Remember the Sand Lizard's protected status and avoid approaching them too closely.

Slow Worm

Anguis fragilis

Length 35cm

| 1 | 2 | 3 | 4 | 5 |

J F M A M J J A S O N D

This animal is a lizard without legs, so resembles a small snake but has a uniformly tubular body with a small head and no obvious neck. It tapers a little at the head and tail end. Its scales are small and flat, giving its body a smooth and polished appearance. Its colour varies from various shades of brown or grey to almost pink-red. Mature males are uniform in colour on the upperside with a mottled belly – some older individuals show faint blue spotting. Females usually have a dark stripe along their backs, spotting on their sides, and a black underside. Slow Worms occur throughout mainland Britain, but are most common in the lowlands. They are often found in gardens.

It is rare to find Slow Worms out in the open. If you do, keep your approach very slow and careful and you may be able to examine the animal closely. Unlike snakes, Slow Worms have eyelids and will frequently blink.

How to find

■ **Timing** Slow Worms emerge from hibernation in March, and mate in May. Young are born in early autumn. They mainly hunt at night.

■ **Habitat** Slow Worms may live anywhere with thick ground vegetation full of invertebrates like beetles, slugs and snails, and with some open, sheltered spots for basking.

■ **Search tips** These animals are often found accidentally, while working in the garden, as they shelter by day underneath stones and logs, and particularly like to hide buried in warm compost heaps. You may see them active in the evenings if you sit outside quietly, but most hunting takes place under deep cover.

WATCHING TIPS

Even if you know you have Slow Worms in your garden, seeing them in action is very difficult, and even more so in the wider countryside. In the garden, you could try creating clearings between known sheltering places, and watch at dusk and dawn. Maintaining a garden to appeal to Slow Worms means avoiding pesticides and leaving plenty of shelter for them – you could try placing squares of old carpet or corrugated metal to act as refugia, and an open compost heap will attract them too. Pet cats can prey heavily on Slow Worms and this is one case where a bell on the collar will not help – avoid encouraging Slow Worms if you have cats that hunt.

Triturus cristatus

Great Crested Newt

Length 16cm

1	2	**3**	4	5

J F M A M J J A S O N D

With his bright colours, 'go-faster' stripe on the side of his tail, and lavish body ornamentation, the male Great Crested Newt in full breeding finery is one of our most beautiful animals.

This very large, handsome newt is widespread in mainland Britain, but has suffered serious and prolonged declines. It is strictly protected from harm and disturbance, and is well-known as a species whose presence at a site can halt building works. The skin on its upperside is dark and warty with fine white speckles. It has an orange-yellow, black-spotted belly, and males in breeding condition have a ragged-edged crest along the length of the back, with an additional crest along the top of the tail (unlike the Smooth Newt's continuous back/upper tail crest). The tail has white 'flashes' on its sides.

How to find

■ **Timing** Great Crested Newts arrive at their breeding ponds at night in March and remain there until early August. After that they live on land, beginning their hibernation in September. Courtship and egg-laying takes place through May and June.

■ **Habitat** These newts prefer fairly large, deep and lushly vegetated ponds, and hunt mainly on the bottom. They spend the winter living under rocks or in other sheltered, damp places.

■ **Search tips** Look out for Great Crested Newts at suitable ponds from April, when males will be courting females. They are more difficult to see later in the summer, as they mainly stay near the bottom, but surface from time to time to take a breath.

WATCHING TIPS

Great Crested Newts can be particularly interesting to watch in spring, when males are courting. The display involves the male positioning himself before the female with a stiffly arched back, dancing from foot to foot and shivering the tail – it is, though, hard to observe in the wild. If your garden has a large, deep pond (or space for one), there is a chance that these newts will colonise. It is of course essential not to disturb these rare, protected animals at any time.

Palmate Newt

Lissotriton helveticus

Length 9cm

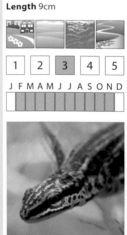

1	2	3	4	5

J F M A M J J A S O N D

Palmate Newts can be found in a wide range of habitats, but perhaps the 'classic' place to find them is in shallow, sunlit pools on open peaty moor and heath, with rare upland dragonflies flickering overhead.

This is our smallest newt – a dainty animal that can breed in very small, shallow pools. It is quite widespread in mainland Britain (absent from Ireland and most Scottish islands) but patchily distributed – it is scarce in central and eastern England, but is the most common newt species in much of Scotland. The name comes from the male's webbed hind feet, a reliable way to distinguish it from the male Smooth Newt (but only in the breeding season). Males also have a less pronounced nuptial crest than Smooth Newt males, and a fine filament at the tail-tip. Females are very like female Smooth Newts, and are best identified by their unspotted throats.

How to find

■ **Timing** Palmate Newts are active from early or mid-spring through to autumn; the active period is shorter in the north of their range. Adults leave their pools in July, and youngsters a month or two later; hibernation begins in late Setpember.

■ **Habitat** This newt is particularly fond of pools on heaths and moorland, which are more acidic than Smooth Newts can tolerate. However, it can also be found alongside Smooth Newts in large and small fish-free ponds in all kinds of habitats, with adjoining well-vegetated terrain offering hiding places for the months spent on land.

■ **Search tips** These newts are shy and prefer to keep within underwater vegetation, but can be easy to spot in smaller and shallower pools, especially around dawn and dusk when they are most active. Keep still and quiet, and watch in spring when the newts will be focused on finding mates. You may find them around the garden outside the breeding season, hiding in leaf-litter – if you do, carefully cover them back up.

WATCHING TIPS

As with other newts, Palmates have a charming courtship display. The male faces the noticeably plumper female, curves his body and tail around towards her and flicks and quivers his tail, to show off its crest and spotted pattern. To attract this species to your garden pond, make sure you have plenty of underwater vegetation and that the pond is kept a fish-free zone. Damp 'cluttered' corners of the garden, with rocks or woodpiles, will provide hiding places for them.

Lissotriton vulgaris

Smooth Newt

Length 10cm

1	2	3	4	5

J F M A M J J A S O N D

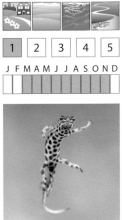

Staring into a wildlife-rich garden pond is a pleasant way to pass the time, and in spring you will probably see Smooth Newts swimming rapidly to the surface for air from time to time.

Also known as the Common Newt, this is our most familiar newt species. In breeding condition, the male has an uneven crest running continuously from behind his head to his tail-tip. His body is grey-green above, orange or yellow below, and is marked with black spots. Females and males outside of the breeding season lack a crest and are drabber in colour. Females are almost unspotted, and very hard to tell from female Palmate Newts. Smooth Newts occur throughout Britain and Ireland, but numbers thin out in the North and West.

How to find

▪ **Timing** Smooth Newts may be seen in ponds from early spring through to mid-summer, after which they live on land and become mostly inactive from mid-autumn. They are most active at dawn and dusk.

▪ **Habitat** These newts may breed in all kinds of still or slow-flowing, well-vegetated waters, but thrive most in smaller ponds with no fish. When not living in water, they hunt or rest in leaf-litter, under logs, and other sheltered places.

▪ **Search tips** Watch any suitable pond on a spring or early summer morning and you'll probably spot a Smooth Newt swimming along or coming up to the surface for a gulp of air eventually. They will dash for cover if alarmed so keep still and quiet, and if the water is clear, you could see some underwater interactions or hunting behaviour. Smooth Newts may also be found in the garden at any time of year; they resemble slow-moving, soft-skinned lizards. Newts found out of water should be left alone – they do not need to be returned to a pond.

WATCHING TIPS

In courtship, the pair sit at the bottom of the pond and the male shimmies and dances around the female, showing off his crest and spots. He then deposits a soft white parcel of sperm, which she sits on to absorb into her cloaca. She wraps each of her fertilised eggs individually in an underwater leaf, and after 10–20 days the tiny, delicate 'newtpoles' emerge, to spend the next couple of months feeding and growing. You can watch all of this behaviour with patience and access to a suitable pond.

Common Frog

Rana temporaria

Length 11cm

1	2	3	4	5

J F M A M J J A S O N D

The Common Frog is more dainty and more contrastingly patterned than the introduced Marsh Frog, which now outnumbers it in some parts of south-east England.

This is our most common and, in many areas, our only frog species. It is a relatively small frog, highly variable in colour but best identified by its dark 'bandit mask', obvious on most individuals, that extends well behind the large eye. Common Frogs are quite abundant throughout Britain and Ireland, but are declining. Garden ponds as a habitat are extremely important for this species, and any carefully designed wildlife pond stands a chance of attracting them.

How to find

■ **Timing** Common Frogs are mostly nocturnal, and live mainly on land – they visit ponds in early and mid-spring to spawn. Spawning time may be as early as January in the mild south-west of England, but a couple of months later in the North. You are most likely to see spawning activity in the early morning, after the frogs have travelled to the pond at night. The tadpoles remain in the water until mid-summer, when they have metamorphosed.

■ **Habitat** Small, fish-free ponds with plenty of surrounding and underwater vegetation are best for Common Frogs to spawn. At other times, they can be found anywhere not too dry, with long ground vegetation and cover in the form of logs, stones or similar, often far from open water.

■ **Search tips** Because Common Frogs usually return (or attempt to return) to their 'home' pond to spawn, seeing them is simple if you know of ponds that they use. Look for them on mornings in early spring after mild nights. You will often find them in the garden when tidying up neglected corners where there are places for them to shelter.

WATCHING TIPS

Spawning Common Frogs are far from discreet and if you have a suitable pond then watching the 'frog orgy' will be an annual wildlife highlight. Males compete to get a female in their grip (amplexus), ready to fertilise her eggs. The resultant balls of spawn take two to four weeks to hatch, and the tadpoles take at least four months to transform into tiny froglets and leave the water. Attract Common Frogs by digging a wildlife pond and allowing them to colonise naturally, rather than moving in spawn from another pond (as this can spread disease). The organisation Froglife offers help and advice: froglife.org.

Bufo bufo

Common Toad

Length 10cm

| 1 | 2 | 3 | 4 | 5 |

J F M A M J J A S O N D

The Common Toad is found throughout Britain (but not Ireland) and, with good views, is easy to tell from the Common Frog. It has very bumpy, warty skin, with prominent, bulging parotoid glands (which secrete poison) behind its very beautiful golden eyes. It also lacks the frog's dark eye-mask. Toads also tend to move with a scrambling walk when disturbed, relying on toxic skin for defence. Although still common, these toads are declining rapidly. Like Common Frogs, they can be helped considerably by gardeners being sympathetic to their needs.

Common Toads are less timid than Common Frogs, and less dependent on damp places – you could encounter them a considerable distance from water, though all will walk to a pond when it's time to breed. You may notice warning road signs with a toad picture, close to busy breeding ponds.

How to find

■ **Timing** Common Toads emerge from their hibernation places in late winter or early spring and head for their spawning ponds. They leave the ponds after spawning and remain active on land (mainly at night) until they hibernate in mid-autumn; they may wander considerable distances. 'Toadpoles' leave the water after completing their metamorphosis in mid-summer.

■ **Habitat** Generally, Common Toads prefer to spawn in larger and deeper water bodies than Common Frogs. At other times they live on land, anywhere with good ground vegetation and some damp and shady hiding places.

■ **Search tips** You'll often find Common Toads when tidying up in the garden, as they shelter in log-piles and under discarded flower pots. If you know of ponds where they spawn, you should see them making their way there at dawn on the first warmer nights in early spring. Road signs with a toad symbol warn of places where they are likely to cross. Toadspawn comes in long strings with eggs in a double row, rather than round masses like frogspawn.

WATCHING TIPS

Toads spawn en masse, making for quite a spectacle. Females often arrive at the spawning pond with a male already 'on board', perhaps with other males trying to dislodge him. Your garden may not be suitable for a deep, toad-friendly pond, but may still attract them with lots of ground-hugging plants and other low-level shelter, and by avoiding all pesticides. If you have resident toads, you might see them hunting if you sit outside at night and keep an eye on the lawn edge or other open patches near cover.

Natterjack Toad

Epidalea calamita

Length 8cm

1	2	3	**4**	5

J F M A M J J A S O N D

This rare toad is smaller, shorter-legged and more boldly patterned than the Common Toad, having rather pale sandy, green or pink skin marked with copious small dark warts, and a yellow stripe down its back. It is famously noisy, and is also a faster mover than the Common Toad, tending to run rather than walk. The habitats it favours are fragile, vulnerable to disturbance and easily damaged by bad weather. It has suffered serious declines as a result, but today strict protection and reintroductions are allowing its numbers to slowly recover. Natterjack Toads can be found on the north-west coast of England and south Scotland, parts of south-west Ireland, and on heathland in southern England – in all, there are about 60 colonies.

This is an attractive small toad with a loud voice.

Super sites

- ★ RSPB The Lodge, Bedfordshire
- ★ Sefton coast dunes, Merseyside
- ★ RSPB Campsfield Marsh
- ★ Holme Dunes, Norfolk
- ★ RSPB Mersehead, Dumfries and Galloway

How to find

■ **Timing** Natterjacks emerge from hibernation as early as March in the south, but, at northern sites, not until early summer – they often emerge after rainstorms. They immediately spawn and leave the water. Adults are nocturnal, sheltering in burrows by day. The toadpoles leave their pools in mid-summer – by this time the pools are often well on the way to drying out.

■ **Habitat** It favours warm, open, sandy habitats – primarily on coastal dune systems, marshes and sandy heathland. It spawns in sun-warmed, shallow, often temporary pools, such as those found in dune slacks.

■ **Search tips** The male's courtship call, a loud rolling croak, carries over a kilometre or more and is a sure sign that spawning activity is underway. Beyond spawning season, the toads are best seen at dawn and dusk. This species is protected from disturbance at all times so you must take a 'wait and see' approach when trying to find it.

WATCHING TIPS

Generally, the best way to see this species is to join a guided walk on a nature reserve that hosts a good population, although you may also be lucky on a dawn or dusk walk. The toads usually dig their own burrows but may use Rabbit or Sand Martin tunnels too; watching these may produce sightings. Its habitat needs are well understood and not difficult to recreate in the right surroundings, making life easier for those working to reintroduce it to parts of its former range.

Rare and non-native reptiles and amphibians

Marsh Frog

| 1 | 2 | 3 | 4 | 5 |

J F M A M J J A S O N D

Wall Lizard

Britain has as many (or more) non-native species as native ones, but their numbers can be very small. They derive from accidental or deliberate introductions. The more easily seen species include Marsh Frogs (well established in wetlands in south-east England), Alpine Newts (found at a few widely separated sites, in wooded areas), Red-eared Terrapins (found on park lakes all around Britain), Aesculapian Snakes (found along the Regent's Canal in London, and in Colwyn Bay, north Wales) and Wall Lizards (present at coastal sites in southern England). The extremely rare Pool Frog is now generally considered to be native to Britain – it occurs very sparingly in east England.

An interesting range of non-native reptiles and amphibians can be found in Britain, but it is pure luck that none so far have proved to be seriously invasive. The problems caused by Cane Toads in Australia show how much harm can be done by introduced species. The deliberate release of any non-native wildlife is illegal.

How to find

■ **Timing** These animals are most active in the warmer months, and many hibernate through winter. Warm, still and sunny days are the best times to look for reptiles like Wall Lizards, while the amphibians are most likely to be seen early or late in the day, in early or mid-spring.

■ **Habitat** This diverse group of animals has similarly diverse habitat needs, though the reptiles all appreciate sunny open spots to bask, and the amphibians all require water for spawning. The 'water frogs' (Marsh, Pool and Edible) spend most of their time in open fresh water through the whole summer, and are quite vocal and visible.

■ **Search tips** Your local herpetology group is your best starting point. Some species are easily located once you are at the right site – Wall Lizards and Marsh Frogs are very easy to see on good-weather days, and some are noisy, such as the Midwife Toad. You will often see Red-eared Terrapins basking on logs or rocks at the water's edge.

WATCHING TIPS

All of these animals are potentially interesting to watch, especially those that are active by day and not shy by nature. The Aesculapian Snake is an expert tree-climber, and the male Midwife Toad carries strings of fertilised eggs wrapped around its legs. A few are potentially harmful to our native wildlife – in particular the Red-eared Terrapin. This is a long-lived species but is not thought to be able to breed regularly in our climate. Sightings of non-native species can be logged with the amphibian and reptile conservation group Froglife – visit froglife.org for details.

Reptiles and amphibians

Britain has rather few species of reptiles and amphibians, but these enigmatic animals hold a great fascination nonetheless. They are also particularly vulnerable to disturbance, so it is important to be careful when seeking them out and watching them, and if you are lucky enough to have any of them resident in your garden, they should be encouraged and cherished.

Special sensitivities

Reptiles and amphibians are ectotherms – reliant on the ambient temperature to get them warm enough to be active. That means they are inactive through the winter out of necessity and are also slow to get going on cooler days in spring and autumn. Reptiles in particular need to spend time in warm places before they can move around at full speed, often basking in direct sunlight. On sweltering days the reverse is true: reptiles cannot safely bask if the air temperature is too high and instead they take cover in shady, cool places.

A warmed-up reptile can move quickly and is very sensitive to ground vibrations. Therefore, if you are out walking in search of snakes or lizards, it's important to tread as softly as you can. Amphibians that are in or

A good 'crop' of **tadpoles** will provide interesting garden wildlife-watching for months as they grow.

alongside water will also sense heavy footsteps and react quickly to a sudden shadow, often swimming into deep water. If possible, approach pools from a direction where you won't cast a shadow. With a careful approach, you can get very close to amphibians and reptiles without alarming them.

A home for herps

The study of reptiles and amphibians is herpetology so these animals are sometimes collectively known as 'herps'. Gardens offer essential habitat for several species of 'herps' – in particular, Slow Worms, Common Frogs, Common Toads and potentially all three species of newts. All of these animals feed on insects and other invertebrates, and all of them need to spend some time living on land, so your garden will need to have good ground cover for them to hide in, and to be a pesticide-free zone to encourage plenty of prey. A compost heap is a warm haven for Slow Worms (and may also attract Grass Snakes), while for the amphibians you can create excellent hiding places with piles of stones, log-piles, bits of broken flower pots and other shelters in shady corners.

Don't stock ponds for amphibians with fish of any kind. Ponds should have gently sloping sides, varying depths and lush native vegetation within the pond and around the margins, to encourage insect life and offer shelter for the young amphibians when they first move out onto land. Avoid the temptation to bring in spawn from other ponds or relocate adult amphibians – this could allow diseases to spread. Natural colonisation may take a while but is worth the wait.

A basking **Adder** is extremely sensitive to disturbance. Approach with care and never get too close.

The **Common Frog** is a species that benefits hugely from sympathetic garden management.

Know the law

All of our native 'herps' are protected by law – it is illegal to deliberately kill them. Several rare species are additionally protected from disturbance of any kind. Always err on the side of caution when dealing with reptiles and amphibians – avoid handling them unless absolutely necessary and be wary of revealing site information about sensitive species to other wildlife watchers, as not everyone is well-intentioned. To report any interesting sightings, and to seek advice on any aspect of amphibian- or reptile-watching or conservation, visit Froglife (froglife.org).

It is tempting to collect or move **frogspawn** around but it's best to leave it well alone, unless moving it is absolutely essential.

Pike

Esox lucius

Length 55cm

| 1 | 2 | 3 | 4 | 5 |

J F M A M J J A S O N D

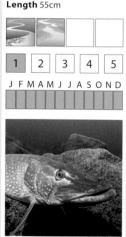

The huge, shovel-shaped head makes the Pike unmistakeable, and that long, almost crocodilian mouth hides the array of sharp, backward-pointing teeth that make this fish such a devastating predator.

This is a large, distinctive fish with fiercely predatory habits, and is common in most kinds of freshwater habitats throughout Britain and Ireland. Although on average an adult is about half a metre long, exceptional individuals grow to more than a metre long. Pikes are usually solitary and can often be seen 'hanging' motionless in the water, waiting for prey – they then attack at lightning speed. They mainly feed on other fish but will take other prey, including amphibians and the young of water birds. They are long, sleek fish with rather small fins, dull greenish-grey in colour with numerous pale spots, and have large wedge-shaped heads with long jaws, the lower jaw extending a little beyond the upper.

How to find

■ **Timing** Pikes may be seen at any time of year but tend to stay in deeper water in winter. They spawn in spring when the water temperature reaches about 9 or 10°C.
■ **Habitat** This species can be found in any fresh water body that holds lots of other fish, and occasionally in brackish water too, but is most common in still or slow-flowing water with plenty of underwater vegetation.
■ **Search tips** Look out for Pikes lurking in shallow water, partly hidden by underwater vegetation. Smaller, immature Pikes prefer more vegetated spots to give them cover from other predators (including bigger Pikes!), while large Pikes may be seen out in the open. They will often wait at narrow points in a water body, where attacking passing fish will be easier. If you are birdwatching and notice a duckling or Moorhen chick suddenly dragged down from underwater, this is almost certainly a Pike attack. In early spring, check warm, shallow lake margins for spawning Pike activity.

WATCHING TIPS

Pike-watching can be rather dull as these fish are fully capable of hanging still in the water for very long spells, but when they do attack prey, their speed is astonishing to behold. Spawning Pikes are often easy to watch as they prefer very shallow water for this, and you are likely to see fins and lengths of back coming to the surface. Pikes are often targetted by fish-hunting birds, such as Cormorants, and will put up a fierce fight in their struggle to escape.

Salmo salar

Salmon

Length 75cm

1	2	3	4	5

J F M A M J J A S O N D

Super sites

★ Falls of Shin, Scottish Highlands
★ Pitlochry Dam Fish Ladder, Scottish Highlands
★ Philiphaugh Salmon Viewing Centre, Borders
★ Stainforth Force (River Ribble), Yorkshire
★ Cenarth Falls, Pembrokeshire

The Salmon or Atlantic Salmon is a magnificent beast and a favourite quarry of anglers – they can grow to more than a metre long. Adults are long, slim, powerful fish, silver-grey with a shining blue and pink lustre, and fine pale and red spots. Young Salmon (parr) are present in our rivers all year round, and you may see them in groups – they are marked with neat, dark, vertical bands that fade as they mature and prepare to make their way to the sea, at somewhere between one and four years old. After another one to four years living at sea, mature Salmon return to their spawning grounds, swimming and leaping upriver. Salmon breed widely on rivers in western and northern Britain.

How to find

■ **Timing** Autumn (October and November) is the time to see adult Salmon heading 'home'. Through late autumn the fish spawn, and many will then die, but a proportion do return to the sea after spawning, though they are more difficult to see as they are much less numerous and are travelling with the river flow.

■ **Habitat** Salmon spawn in clear, cold rivers with gravelly bottoms. The young Salmon remain in these rivers over the time it takes them to mature. Migrating Salmon are easiest to see on fast-flowing, steep stretches of river where they have to leap to make progress.

■ **Search tips** There are several sites where migrating Salmon can reliably be seen at the right time of year. Wait and watch and you'll soon be rewarded with good views. Look for young Salmon in sheltered river edges where the water flows more slowly.

WATCHING TIPS

The upstream migration of Salmon is wonderful to watch – truly one of the great wildlife experiences. The best views are often right after heavy rainfall, which increases the river flow rate and forces the fish to work harder. Find a good vantage point and watch quietly. The fish are very intent on their task but may still be alarmed if they see or feel sudden movement nearby. Photographers may have most success by pre-focusing manually on a particular spot where fish have already been seen jumping, and firing off shots as the next fish approaches.

Brown Trout

Salmo trutta

Length 60cm

1	2	3	4	5

J F M A M J J A S O N D

This fish is a close relative of the Salmon, but is usually a little smaller. It is an attractive sleek-bodied fish with silver and gold colouration, marked all over with dark spots. The form of Brown Trout known as 'sea trout' is silver-coloured, and spends its adult life at sea, only returning inland to spawn – it is typically larger than the form that does not have a sea-dwelling stage. This fish is widespread throughout Britain and Ireland, anywhere where there is clean and clear cold water with plenty of prey for it. The colourful Rainbow Trout is not native to Britain but has been widely introduced here.

Fly-fishing for Brown and Sea Trout is permitted from late spring through to early or mid-autumn (exact dates vary by region). The fish are easier to see (as well as catch) during these months.

How to find

■ **Timing** Brown Trout may be seen at any time of year, but are less active and stay in deeper water in the coldest months. They return to their birthplace streams to spawn in late autumn, and spawning activity continues through to December.

■ **Habitat** This fish, like the Salmon, prefers clear, cold water with a gravelly substrate, and, while it spawns in streams, it may be found in still waters at other times.

■ **Search tips** If you stand on a bridge overlooking a suitable stream, river or stretch of lake, you may well spot shoals of small Brown Trout, but mature adults at spawning time tend to be territorial, especially in rivers. Early evenings in September are good times to look for Brown Trout as they will be feeding avidly, in preparation for spawning. When there is a large hatch of insects such as mayflies, you are likely to see trout coming to the surface to take them.

WATCHING TIPS

The spawning behaviour of trout is interesting. Females arrive at their natal stream and dig out a hollow in the gravel substrate (a 'redd'). Once the redd is complete and water temperature is right (7–9°C), the female will enter, along with her chosen mate, and release her eggs for him to fertilise. The female then covers up the eggs with the gravel she had moved aside, and the eggs will remain there through winter, hatching the following spring. Spawning takes place in shallow water so can easily be observed if you are lucky with your timing.

Gasterosteus aculeatus,
Pungitius pungitius

Three-spined and Nine-spined Sticklebacks

Length 4cm

1	2	3	4	5

J F M A M J J A S O N D

Three-spined Stickleback

Nine-spined Stickleback

Sticklebacks are very small fish – the classic 'tiddlers' of streams and ponds – and have spines on their backs in front of the dorsal fin. The Three-spined Stickleback is a very common and widespread species, while the Nine-spined is also widespread but scarcer. Although the number of spines present on their backs is variable in both species, there is no overlap – Three-spined has two to four spines, while Nine-spined has nine to eleven. Males of these little fish become very colourful at spawning time and are interesting to watch, with elaborate courtship behaviour.

If your garden pond is well-established, you may find that it already has some sticklebacks. While their presence will discourage amphibians, a solution is to dig a second, fish-free wildlife pond, as a haven for frogs and newts.

How to find

■ **Timing** Sticklebacks may be seen at any time of year, but are more active and more likely to be seen well in the warmer months. Courtship and spawning takes place in late spring through to early summer.

■ **Habitat** Both species can be found in both fresh and brackish waters, from lakes to small ponds and in streams and slow-flowing rivers. They require underwater vegetation for their nests.

■ **Search tips** Sticklebacks (especially Three-spined) are likely to be present in many, if not most, small ponds, and if you take part in a pond-dipping activity at a nature reserve you stand a good chance of seeing one at close quarters. Watching at a clear stream or pond may produce good views, if you are patient. It is not advisable to introduce them to your garden wildlife pond though, as they will prey on amphibian larvae.

WATCHING TIPS

Watching stickleback courtship is fascinating, though difficult. Males develop vivid red and blue coloration at this time of year and establish a defined territory in shallow water. Here they build a nest from underwater debris, and then attract females into it by performing a quivering display dance back and forth, displaying their bright colours. Females enter the nest and lay their eggs, which the male fertilises, and then guards until they hatch about a week later.

Common Carp

Cyprinus carpio

Length 70cm

1	2	3	4	5

J F M A M J J A S O N D

This is one of the most easily seen freshwater fish in Britain. It is native to mainland Europe and was introduced to Britain in the Middle Ages as a food fish, though today it is rarely eaten, but caught (and released) for sport. It is a large, thickset, deep-bodied fish, with prominent and attractive even-sized scales over its body. In its domestic forms, the 'leather' carp has no obvious scales and the 'mirror' carp shows very large, uneven, shiny scales. These forms tend to be larger, and can grow to more than a metre long. Common Carp are found in large, lowland lakes throughout England, and more sparsely in Wales, Ireland and Scotland.

Common Carp rarely appear in a hurry – look out for their broad backs moving through clumps of water-lily leaves. At some lakes, they are used to being fed and will gather at the surface near the bank in the hope of a thrown bread crust.

How to find

■ **Timing** As with most other freshwater fish, Common Carp are easiest to see in the warmer months. They enjoy sunshine and bask near the surface on sunny days. Spawning takes place from mid-spring to early summer.

■ **Habitat** These carp are most likely to be found in large, lush water-bodies with plenty of underwater and emergent vegetation. Fishing lakes are often stocked with large numbers of Common Carp.

■ **Search tips** You will often notice Common Carp swimming close to the surface in a stately manner, breaking the surface with their dorsal fin and upper back. In some places, they are accustomed to being fed by human visitors and will gather at the bank side or under bridges. If you visit a lake and notice anglers set up with bivouacs and rods with 'bite alarms', that is a sign that there are big carp present.

WATCHING TIPS

These large, handsome fish are easy to watch on summer days as they nose their way through the water lilies and bask and cruise near the surface. They are bottom-feeders, eating snails that they find on the muddy lake bed, but will come to the surface to feed at times, or to gulp air. When spawning they become considerably more lively and fearless and males battle vigorously for female attention in the shallows, sometimes even beaching themselves through their energetic thrashings. Males in breeding condition can be recognised by the white tubercles on their heads, while gravid females become even more rotund than usual.

Scardinius erythrophthalmus, Rutilus rutilus

Rudd and Roach

Roach

Length 45cm

1	2	3	4	5

J F M A M J J A S O N D

Rudd

These two common freshwater fish are found in the same kinds of habitats and are superficially very similar. Both are relatively small, silver-coloured fish, their bodies somewhat flattened on the vertical plane, with reddish or bright red fins. The Rudd has yellow eyes and an upturned mouth, indicating its preference for feeding at the surface, while the Roach has red eyes and a more horizontal mouth for all-level feeding; the Rudd also tends to be deeper-bodied than the Roach. Both are fairly easy to find and to watch, and occur widely in Britain and Ireland.

A shoal of silvery freshwater fish sporting red fins are likely to be Rudd or Roach. Check the eye colour and mouth shape to identify them – close-focusing binoculars are helpful here.

How to find

■ **Timing** These fish are most active in the summer months and most likely to enter shallower water at this time as well. They spawn from early spring to early summer.

■ **Habitat** Both species are adaptable and occur in all kinds of reasonably large fresh waters. They prefer still or slow-flowing waters with plenty of underwater vegetation. The Roach is particularly tolerant of poor water conditions and may be the only fish present in nutrient-poor lakes – it can also tolerate brackish water.

■ **Search tips** These fish are often spotted in shoals in shallow water, drawing attention with their striking red fins. Look out for them loitering near the surface in sun-warmed waters, or moving among underwater plants. When searching, move slowly and step softly, keep quiet and try to avoid casting your shadow across the water as they are easily alarmed.

WATCHING TIPS

Roach and Rudd are among the easier of the smaller freshwater fish to watch. A riverside picnic in a good spot overlooking some open water will give you the opportunity to watch their shoaling behaviour at length, and in spring you may see spawning activity. This can be frenetic, with males chasing gravid females and sometimes jumping clear of the water. Although they don't grow very large, these fish can live for 15 years or more. However, only a tiny proportion of fry will reach adulthood, as each female can produce 100,000 eggs each time she spawns.

Perch

Perca fluviatilis

Length 25cm

1	2	3	4	5

J F M A M J J A S O N D

This is a particularly attractive freshwater fish. When mature it is rather deep-bodied with a steeply rising 'hump' back. Its body is shining silver-green with large fins (the lower ones red-tinged), and has a row of bold, vertical, dark stripes down its sides. The dorsal fin is long and tall, with spiny rays, and a bold black spot at its back end. Perch are common in clean, well-oxygenated lakes and slow-flowing rivers, with plenty of underwater vegetation and other submerged hiding places. They are shoaling fish but are also fierce predators, adults taking many other small fish. Perch are present throughout Britain and have been introduced to Ireland.

The rays in a Perch's dorsal fin end in sharp points, which make it a tricky fish to handle for anglers and wild predators alike. Despite this hazard, the Perch is a popular quarry of grebes and cormorants.

How to find

■ **Timing** You're most likely to see shoals of Perch between spring and autumn, when they will spend more time near the surface and in shallower water. As with most freshwater fish, in winter they will retreat to deeper water where the temperature is more stable. They spawn in mid-spring.

■ **Habitat** Although its scientific name relates to rivers, the Perch is found in lakes and ponds with good water quality, as well as in slow-flowing rivers. It requires plenty of underwater vegetation.

■ **Search tips** This species is a favourite prey of fish-catching birds and you may stand more chance of seeing one in the bill of a Great Crested Grebe than in the water. It is alert and wary so you will need to keep still and quiet when looking for it, but it will often be found close to the shore.

WATCHING TIPS

Perch will form shoals with other similar-sized fish species as well as other Perch in their age class but are efficient hunters of any fish smaller than themselves. Small young Perch are hard to watch as they keep close to cover for safety, but larger fish spend more time in the open. At spawning time, females produce long strands of sticky eggs which they wrap around underwater plants – males then follow to fertilise the eggs. Those Perch that live longest can grow much larger than average (60cm or more) and may survive well into their twenties.

Other freshwater fish

Barbel

1	2	3	4	5

J F M A M J J A S O N D

Minnow

Many other species of freshwater fish occur in Britain, including Bream, Tench and Chub, and the tiny Minnow. Fish found mainly or exclusively in rivers include Grayling, River Lamprey and the beautiful Barbel. It is obviously difficult to see fish in their natural environment, especiallybottom-feeding species like Bream which prefer deep and turbid waters. However, small fish that inhabit clear shallow streams may be seen more easily, including scarcer species such as Brook Lampreys, Stone Loaches and Bullheads. The European Eel, a once abundant, strongly migratory species that is now classed as Critically Endangered after severe declines, may even be seen on land, as it can move snake-like over wet grass.

Chatting to anglers is a good way of finding out where you might observe particular fish species around your local waterways. Even the smallest species have their fans within the angling community.

How to find

■ **Timing** All of our freshwater fish are easier to see in the warmer months, and many are most active early and late in the day. Spawning times vary between species – the majority spawn in spring but a few spawn in autumn, their eggs not hatching until the following spring.

■ **Habitat** All kinds of still and flowing water can offer opportunities to find different species of freshwater fish. Diversity and numbers will be highest in large, well-vegetated lowland lakes, but clearer waters offer a better chance of good views. Clear, cold rivers and streams are likely to support fish.

■ **Search tips** When on the lookout for fish, dress in dull colours and try to avoid casting your shadow over the water. Pause often and look around clumps of underwater vegetation where fish can hide. Take some time to 'get your eye in' for the subtle movements of small fish. Watching from a bridge or a weir is a good idea. You might also try using polarising sunglasses, which cut out some of the glare from the water's surface.

WATCHING TIPS

Fish-watching is often tricky but can be rewarding. There is also the potential to make some real 'citizen science' discoveries in less well-studied areas. If your interest grows, you could also try underwater film recording. Confident swimmers could also join a local outdoor swimming or diving group, and (after appropriate training) you'll have the chance to get up close and personal with wild fish on their terms. Pond-dipping may turn up some small fish alongside the more expected aquatic invertebrates, and using an aquascope or bathyscope is a great way to watch underwater life while staying dry yourself.

Rockpool fish

Butterfish

1	2	3	4	5

J F M A M J J A S O N D

Tompot Blenny

Rockpooling offers the chance to see a great variety of animals at close quarters. For many rockpoolers, it is the little fish that are the biggest draw, and among them are some spectacularly beautiful and bizarre-looking creatures. They include Common and Rock Gobies, sleek and eel-like rocklings, blennies with their beautifully camouflaged patterns and tall, spiny fins, and colourful wrasses and Butterfish. You'll find the best variety in large rockpools closest to the shore, which are only exposed at the lowest tides – there is even a chance of finding seahorses in such pools.

Fish that are adapted to live in rockpools are often expert at hide-and-seek. Some are long and slim to slip into tiny spaces between rocks, while others have camouflage that involves their whole body shape as well as their colouration.

How to find

■ **Timing** When searching for rockpool fish, the lower the tide, the better. Consult tide tables to plan your trips. Fish activity is likely to be highest in the early morning and the evening.

■ **Habitat** The best rockpools are large and deep, with lots of crevices, seaweed and loose stones on the bottom to offer hiding places for fish. Areas with extensive stretches of rocky shores will hold more different species than more isolated rocky patches on otherwise extensive sandy or shingle beaches.

■ **Search tips** First, watch your pool for signs of life and movement. You can then try gently turning over stones and moving clumps of seaweed aside. Always gently place moved objects back to exactly where they were resting before. Shine a torch into dark crevices to see what's hiding within. When exploring rockpools, step slowly and carefully as the surface may be slippery, and take care not to dislodge limpets and other animals attached to the rock. Some rockpools are large enough to swim and snorkel in – keep your movements slow and gentle as you explore.

WATCHING TIPS

Rockpool fish tend to be sit-and-wait hunters, using camouflage to avoid detection as they rest on the bottom, waiting for prey to come along. Many spend hours hiding well out of sight. However, with patience you can see some interesting sights, such as beautiful male Corkwing Wrasses creating their seaweed nests, or fanning the eggs with their tails to keep them well oxygenated. Some rockpool fish, such as the Tompot Blenny, are inquisitive and may emerge from their shelters to inspect you if you keep your movements slow and careful.

Cetorhinus maximus

Basking Shark

Length 6m

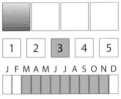

1	2	**3**	4	5

J F M A M J J A S O N D

Observing marine fish in their natural environment is not generally possible for the average wildlife-watcher without considerable investment in specialist equipment and training, and so is beyond the scope of this book. However, a couple of sea fishes can be seen and watched fairly easily around the British coast without needing to get into the water, and one of these is the Basking Shark, the second largest fish in the world. This plankton-eating shark roams temperate seas all around the world, but often comes quite close inshore and may be seen from land, as well as boats – its vast size and slow movements make it unmistakeable. The best areas to encounter it are off south-west and western coasts.

That huge mouth looks imposing indeed, but the Basking Shark has no interest in swallowing snorkelers.

Super sites

★ South-western headlands, and pelagic wildlife-watching trips from Cornwall
★ West coast of Scotland

How to find

■ **Timing** Most Basking Shark sightings occur between May and September, with the season beginning earlier in the South. They are most likely to be near the surface (and are easier to see) in calm waters.

■ **Habitat** Look for Basking Sharks in areas where the continental shelf is close to the land – the upwellings bring its plankton food close to the surface.

■ **Search tips** If watching from land, a headland is ideal – find an elevated position with the light behind you and a wide field of view across the surrounding sea. Scan often, looking for the typical tapering shark outline near the surface. Boat trips into suitable seas can also be good for seeing Basking Sharks – in this case, you will probably first spot the tall, straight dorsal fin showing clear of the water, and perhaps the long broad snout as the animal feeds.

WATCHING TIPS

An encounter with this marine giant is unforgettable. Basking Sharks are usually solitary but sometimes turn up in small groups, and then you might see social interactions, such as (slow-motion) nose-to-tail chases. On rare occasions, Basking Sharks even breach fully clear of the water. Their typical behaviour, though, is extremely sedate, and they are not aggressive in the least, so close and prolonged views from boats are often possible.

Sunfish

Mola mola

Length 2m

1	2	3	4	5

J F M A M J J A S O N D

The side-floating habit of the Ocean Sunfish makes it look like a fishy corpse at first glance.

This very large marine fish spends much time near the surface, and so can regularly be seen from land or boats. It is a bizarre-looking animal, with a flattened, almost circular body, no true tail-fin, but elongated dorsal and anal fins that project at right angles to the rear of the body and act together as a rudder. The largest Sunfish can exceed 1,000kg, heavier than any other bony fish. When not feeding in the depths it often comes to the surface and lies on its side. Sunfish prefer tropical and warmer temperate waters, but regularly wander far enough north to be seen in the Atlantic around south-west England. They feed mainly on jellyfish, but also take squid, juvenile fish and other slow-moving marine animals.

Super sites

* South-western headlands, and pelagic wildlife-watching trips from Cornwall
* West coast, Scotland

How to find

- **Timing** Look for Sunfish on warm, calm days in summertime.
- **Habitat** Sunfish hunt in deep waters, so are most likely to be seen close to the continental shelf, in seas with abundant marine life.
- **Search tips** This fish is large enough to be seen from land at a considerable distance if the sea is calm and light conditions are good. Sightings are most likely from a headland with the light behind you and are easier in the morning and evening than midday. When looking for sunfish from a boat, you may first see the narrow tip of the dorsal or anal fin breaking the surface. If you can only see a fin, you can distinguish it from a shark's by the way it moves: up-and-down, rather than side-to-side, when the fish is on the move.

WATCHING TIPS

When you see a Sunfish at the surface, it will probably be doing very little beyond just lying on its side. This behaviour is not fully understood and may have multiple functions, the most widely accepted that it is done to regulate heat, but the fish may also be inviting seabirds to remove parasites from its scales. Sunfish can breach clear of the water, but this is very rare. However, even when a Sunfish is just resting motionless at the surface, having a close look at this extraordinary deep-sea creature is exciting.

Aglais urticae

Small Tortoiseshell

Wingspan 5.5cm

| 1 | 2 | 3 | 4 | 5 |

J F M A M J J A S O N D

This colourful butterfly is one of the group known as 'aristocrats', which make up part of the family Nymphalidae. The Small Tortoiseshell is no longer as ubiquitous as it once was, but can still be seen almost everywhere in Britain and Ireland, and is a strong flyer so can turn up well away from ideal habitat. The upperside is bright orange with dark and pale yellow markings, and a border of blue spots on a dark background on the hindwings. The underside is mottled dull brown, providing camouflage. Small Tortoiseshells are attracted to nectar-rich garden flowers, and often hibernate inside garden buildings.

The drab underside of a Small Tortoiseshell hints at its bright upperside pattern, while providing camouflage that will help it escape the attention of predators through its long months of hibernation.

How to find

■ **Timing** These butterflies hibernate as adults, but may fly on very mild winter days. They emerge fully from hibernation in March or April and breed through spring, producing a first generation which is on the wing in mid-summer. A second generation emerges in early autumn, then hibernates from mid-autumn. The generations overlap, so Small Tortoiseshells are on the wing in varying numbers all through the warmer months.

■ **Habitat** The food plant is Common Nettle, ideally patches with new growth, in full sunshine. Nettle clumps on woodland edges and along hedgerows often provide perfect conditions for egg-laying. The adults visit all habitats with nectar-rich flowers – in late summer they are also attracted to windfall fruit.

■ **Search tips** As soon as the spring air reaches about 16°C in at least some sheltered spots, look for this colourful butterfly basking or seeking nettle clumps. The last generation of the year will feed constantly, fuelling up for their long hibernation. Check outbuildings for hibernating adults.

WATCHING TIPS

The classic butterfly-attracting garden flowers, such as buddleia, appeal to this species. Blackthorn is an important nectar source at the start of the flight season, and ivy at the end. Courtship behaviour is best observed in early spring. If you find these butterflies hibernating in your home, gently move them to a sheltered spot with easy access to the open – they are likely to desiccate in heated rooms. Large and Yellow-legged Tortoiseshells are occasionally recorded in Britain – both resemble Small Tortoiseshells but are bigger and duller-coloured. Check and photograph any odd-looking tortoiseshell butterflies you encounter.

Peacock

Aglais io

Wingspan 6.7cm

1	2	3	4	5

J F M A M J J A S O N D

This gorgeous butterfly, part of the 'aristocrat' group, is unmistakeable when seen well. The underside is almost black, giving a dark appearance in flight. The upperside is bright velvety red with dark wing margins, and very large, colourful black-edged eye markings at the upper edge of each wing. The caterpillar is also conspicuous with its black, white-speckled and very spiny body. The Peacock is common throughout Britain and Ireland, though scarcer in uplands in the far north of Scotland.

Peacock caterpillars feed communally on clumps of Common Nettle. Their spiny skins make them unappealing prey, so they make no real attempt to conceal themselves.

How to find

■ **Timing** This species hibernates as an adult, emerging on warm days in early spring. This generation breeds through spring, and the resultant larvae will be seen through summer. The old adults have mostly disappeared by the start of June. The new generation of adults flies from July to mid-autumn when hibernation begins.

■ **Habitat** Peacocks lay their eggs on Common Nettles, preferring new growth but in slightly less shaded conditions than Small Tortoiseshells choose. Adults wander widely, visiting nectar-rich flowers in gardens, parks, woodland edges, along hedgerows and around lakes. They hibernate in sheltered places, including inside buildings.

■ **Search tips** Peacocks emerging from hibernation in spring are restless and keen to breed – often all you see is a large dark butterfly flying past very strongly and quickly, but they may pause to bask on warm ground. The caterpillars are conspicuous through early summer. Fresh adults in late summer and autumn are only interested in feeding and are easy to find on sunny days, visiting buddleia, bramble, ivy and other nectar-rich flowers.

WATCHING TIPS

Peacocks enjoy basking with wings spread, giving the opportunity for close views and photos. A garden appealing to Peacocks will have plenty of nectar-rich flowers that, between them, flower from March to October, some bare ground for basking and a partly sunlit nettle patch, which may tempt them to lay eggs. If a Peacock tries to hibernate in your house, you should move it somewhere sheltered outdoors. An unheated outbuilding with easy access to the open air is an ideal choice.

Vanessa atalanta

Red Admiral

Wingspan 7.2cm

| 1 | 2 | 3 | 4 | 5 |

J F M A M J J A S O N D

This familiar large butterfly formerly occurred in Britain only as a migrant, but since the late 20th century has been successfully overwintering in southern areas. The resident population is still greatly boosted through summer by arrivals from the continent. It has a dark upperside with a bold red band across all wings, and white spots in the forewing wingtips. The underside is dark and mottled for camouflage, though subdued red and white markings are present on the forewing undersides. It is found throughout the British Isles but is most numerous in the South.

The Red Admiral's species name, *atalanta*, references an Ancient Greek athlete who lost her race because she was distracted by an apple on the ground – and like her, this butterfly adores windfall fruit.

How to find

■ **Timing** Hibernating adults emerge in March, though they may fly on mild days in winter. They breed through spring and give rise to a new generation that emerges in July, but because immigrants constantly arrive from mid-spring, there is not the same noticeable quiet period in early summer that is seen with Peacocks and Small Tortoiseshells. They feed avidly through late summer and autumn, and are regularly seen on mild days in November and December.

■ **Habitat** Red Admirals are strong flyers with nomadic habits, so may turn up anywhere with nectar-rich flowers. They are also strongly attracted to windfall apples and other soft fruit. The usual food plant is Common Nettle.

■ **Search tips** This is a conspicuous butterfly, easily seen visiting buddleia or bramble flowers in summer, and frequently basking on bare ground. The caterpillars make themselves a 'feeding tent' by drawing together the sides of a nettle leaf with silk – these tents are easy to spot. The full-grown caterpillar pupates within a similar tent.

WATCHING TIPS

Even a tiny city garden stands a good chance of attracting Red Admirals in late summer, as long as there are flowers from which to feed. Red Admirals also bask at length, resting with wings wide open in all but the hottest weather, so are a gift for photographers. Numbers seen each year vary depending on how many migrants arrive from the continent. Some of these may make a return migration, after fuelling up. Ivy flowers are a particularly important nectar source late in the year.

Painted Lady

Vanessa cardui

Wingspan 7cm

1	2	3	4	5

J F M A M J J A S O N D

This species is our best-known migrant butterfly, arriving here in varying numbers – sometimes the influx is enormous. It is resident in north Africa and around the Mediterranean, but migrates northwards from spring, and may be seen anywhere in Britain and Ireland. It breeds here but cannot survive our winter. However, the new generation of adults may undertake a southward return migration. It has a salmon-pink upperside with black markings and white spots in the wingtips – the underside is mottled in shades of light brown. On the wing, it looks rather pale and it has a fast, powerful flight.

Painted Ladies will spend a long time working their way up and down a buddleia bloom, but can be flighty and will be gone in an instant if disturbed. In flight they look light orange-pink, paler than other 'aristocrat' species and less fiery-orange than the fritillaries.

How to find

■ **Timing** The first Painted Ladies arrive as early as April, and continue to turn up through summer. The offspring of the first arrivals will emerge as adults from early July. Sightings dwindle through early autumn and few are seen after September.

■ **Habitat** Painted Ladies lay their eggs on various species of thistles, and also on Common Nettles. The adults roam widely in search of food and mates, and may visit anywhere with flowering plants – warm, sunny, open ground is particularly appealing to them. Because they are migrants from further south, they can be numerous on southern and eastern coasts.

■ **Search tips** In the best years, such as 2009 when an estimated 11 million Painted Ladies arrived here through spring, they can vastly outnumber other species at many sites. Meadows with stands of thistles, and sunny gardens with buddleia, are good hunting grounds. They often rest on bare ground where they can be surprisingly inconspicuous, taking flight at your approach and disappearing very quickly.

WATCHING TIPS

In an exceptional migration year, it is worth taking a trip to the coast to witness mass arrivals. Keep an eye on social media – it will usually be evident by May whether a 'big year' is on the cards. A garden full of butterfly-attracting flowers can bring in Painted Ladies all through summer. The return migration after such a year should be even more considerable but the butterflies travel back at very high altitudes so are difficult to observe. In bumper years for Painted Ladies, it is worth remembering that other migratory insects may also turn up in increased numbers.

Polygonia c-album

Comma

Wingspan 6cm

| 1 | 2 | 3 | 4 | 5 |

J F M A M J J A S O N D

In hot weather, a Comma will keep its wings closed to avoid overheating. Males rest on a vantage point with good visibility, ready to attack any intruders that wander into their territory.

This butterfly has a distinctive, very jagged wing outline. It otherwise recalls a fritillary or a less colourful Small Tortoiseshell, with orange wings marked with dark brown spots. The underside is plain brown (distinctly paler in the non-hibernating *hutchinsoni* form), with a small white c-shaped marking on the hindwing. The Comma is a common and increasing species in England and Wales, and has spread to southern Scotland – it is, however, absent from Ireland. It is a strong flyer but less inclined to wander than the other colourful 'aristocrat' species.

How to find

■ **Timing** Commas emerge from hibernation in early spring. These individuals produce a mid-summer generation, which flies from June or July. Most of these are 'normal' Commas, which do not breed in their birth year but hibernate in autumn. However, a variable proportion of these are of the lighter form *hutchinsoni*. These individuals breed rather than hibernate, producing an additional late-summer generation that goes on to hibernate. Commas can be seen from March through to October and may fly on mild winter days.

■ **Habitat** This is mainly a woodland species, but also visits larger gardens, parkland and other areas with flowers and a mixture of tree cover and sunny open areas. In late summer, it is attracted to orchards to feed on windfall fruit. The main foodplant is Common Nettle.

■ **Search tips** Commas are eye-catching with their bright colours, and bask in sunlight. Males are territorial and sit on a prominent, well-lit perch, flying out to intercept other passing insects and returning to the same spot.

WATCHING TIPS

Commas are feisty, energetic butterflies that are engaging to watch, particularly the males in territorial mode. The proportion of pale *hutchinsoni* individuals that appear in summer seems to be dictated by spring weather, with more appearing in years when conditions are good. This phenomenon has only recently been described and there is much the amateur lepidopterist can contribute to continued study, so consider keeping track of the forms you see and passing the details to your local butterfly recorder.

White Admiral

Limenitis camilla

Wingspan 6cm

1	2	3	4	5

J F M A M J J A S O N D

One of the less colourful 'aristocrat' butterflies, this is nonetheless an extremely beautiful insect with a strikingly elegant gliding flight. It is found in southern, eastern and central England, and has increased and spread considerably over the last few decades. The upperside is velvety black with a broad white band across all wings. The underside is bright tawny-orange, also marked with a white band as well as small black spots. The body is black on the upperside and white on the underside. It could be mistaken for a female Purple Emperor, but is smaller with rounder wings, and does not show the large eye marking on the underside forewing.

The neat, broad white band across both the blackish upperside and golden-brown underside makes this an unmistakable species. A side view shows that the head is dark on top and white underneath.

How to find

■ **Timing** White Admiral adults emerge from the second half of June and are on the wing into early August. They overwinter as larvae. The best time to see them is from mid-morning to mid-afternoon on still, sunny days.

■ **Habitat** This species needs woodland with shady, tangly patches where plenty of honeysuckle (the larval foodplant) grows, and some sunny rides, edges or clearings with brambles and other nectar-rich flowers.

■ **Search tips** Take a slow stroll through sunny woodland, concentrating on the widest glades and rides, and checking patches of flowering brambles. You stand a very good chance of seeing White Admirals feeding from bramble flowers, and as you walk along you may see them patrolling the rides, flying at head-height or above with their very distinctive gliding action, and settling from time to time on high vantage points. They may also come down to ground level to feed on nutrients in muddy patches or from animal droppings.

WATCHING TIPS

White Admirals are delightful to watch on the wing – no other British butterfly moves with quite the same grace. Also look out for females seeking egg-laying sites with a much slower and more fluttering flight. Although White Admirals fly well into mid-summer, it is worth seeking them out at the very start of the flight season, as their wings become worn and tattered very quickly, perhaps because of their fondness for visiting thorny bramble flowers.

Apatura iris

Purple Emperor

Wingspan 8cm

1	2	3	4	5

J F M A M J J A S O N D

This glorious butterfly fully deserves its noble epithet.

This spectacular insect is perhaps the most coveted of all British butterflies, particularly the male with his magnificent violet-glossed wings. It is a rare species and can be difficult to see as it conducts much of its business in the treetops. The Purple Emperor is very large, with long, pointed forewings. The upperside is dark with a narrow, broken white band across all wings, and males' wings reflect a purple sheen in the light. The underside is brown, with a broad white band across the hindwing, and some white markings and a large brown, black and white eye marking on the forewing. It is restricted to southern central England, with the most sites in Surrey, Hampshire and surrounding counties.

Super sites

★ Alice Holt Forest, Hampshire
★ Bentley Wood, Hampshire/Wiltshire
★ Bookham Common, Surrey
★ Knepp Castle Estate, West Sussex

How to find

■ **Timing** Adult Purple Emperors fly between early June and mid-July. They spend the middle of the day engaged in territorial and courtship activity around the tops of tall 'master trees' on hilltops, but you stand a better chance of closer views in the morning or evening when they are feeding.

■ **Habitat** Most Purple Emperor colonies are in extensive mature oak woodland, but other kinds of deciduous woods can support them too. The larval food plant is sallow.

■ **Search tips** Locating and watching a 'master tree' is the surest way of seeing Purple Emperors, but it is also worth keeping watch early at muddy puddles or piles of animal droppings, as males come down to take nutrients before their work begins. They will return to such places in the evenings.

WATCHING TIPS

The courtship behaviour of this butterfly is well-documented and can be observed via binoculars or even a telescope. Males battle in mid-air to secure the top spot in the canopy of the 'master tree' and rush out to intercept any passing females. Mating takes place in the treetop, but usually will be out of view. If the female has already mated, though, she drops down to ground level to shake off her admirers. Males visiting nutrient sources on the ground, most typically first thing in the morning, may spend as long as an hour in one spot, allowing for excellent views.

Pearl-bordered and Small Pearl-bordered Fritillaries

Boloria euphrosyne, B. selene

Wingspan PBF 4.5cm
Small PBF 4.1cm

1	2	3	4	5

J F M A M J J A S O N D

Pearl-bordered Fritillary

Small Pearl-bordered Fritillary

These two small fritillaries are closely related and very similar-looking – to compound matters, their flight seasons overlap and at some sites both species can be seen. Both have tawny-orange uppersides with black bars and spots. The underside pattern is more strongly contrasting in Small Pearl-bordered, with more white spots. Pearl-bordered is rarer, occurring very patchily in south England, the Welsh borders, north-west England and the Highlands, also west Ireland. Small Pearl-bordered has a similar but more extensive and continuous range, including most of Wales and Scotland, but is absent from Ireland. These fritillaries are much weaker fliers than the larger species, and (once found) are easy to watch.

These fritillaries are best sought out in more remote northern and western areas with unspoiled, sheltered meadows and woodland clearings. In such areas, they may even be garden visitors. They somewhat resemble miniature versions of Dark Green and High Brown Fritillaries.

How to find

■ **Timing** Pearl-bordered Fritillaries fly from mid-April to mid-May (appearing earlier in the south of their range). The Small Pearl-bordered flies from early May to mid-June, again emerging earlier in the South. Both species may produce a small second brood in August.

■ **Habitat** Both species prefer woodlands with open sunny areas and damp patches that encourage their foodplant (dog violet) to grow prolifically, but they may also be found in more open habitats, such as moorland and heathland edges.

■ **Search tips** These butterflies are conspicuous as they warm up in the first morning sunlight, with wings wide open. Later on, they become active and spend much of their time feeding from low-growing flowers. Search slowly in sheltered flowery places along glades and woodland edges. You will need to see the underside to make a definite identification, but butterflies that are basking to warm up will usually permit close inspection.

WATCHING TIPS

Their rather slow fluttering flight and preference for feeding at ground level makes these butterflies easy to watch. Males fly about in the middle of the day in search of females, dropping down to inspect any object that is about the right colour. When he finds an unmated female, the two may rise up to a higher perch to mate, but will rest in full view, while females seeking egg-laying sites make a slow exploration of ground-level vegetation. Less powerful on the wing than larger fritillaries, they both thrive best in sheltered places such as woodland clearings, and will take shelter on windy days.

Argynnis aglaja, A. adippe

Dark Green and High Brown Fritillaries

High Brown Fritillary

Wingspan 6cm

1	2	3	4	5

J F M A M J J A S O N D

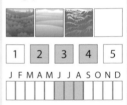
Dark Green Fritillary

These two species, much like the pearl-bordered fritillaries, are a closely related pair with many things in common. They are large, strong-flying butterflies, with fiery orange, black-spotted uppersides, and you'll need to inspect the underside pattern for a certain identification – High Brown has a row of red-circled white spots near the hindwing border, which Dark Green lacks. Dark Green is far more widespread, with colonies throughout Britain and Ireland (though is much more common in the West than the East). High Brown, once also fairly common, is now very rare, with scattered colonies in Devon, Somerset, west Lancashire and parts of Wales.

To identify these species, you need a clear view of the underside hindwing.

Super sites

High Brown Fritillary
★ Dunsford Meadow, Devon
★ Aish Tor, Devon
★ Arnside Knott, Lancashire
★ Gait Burrows, Lancashire
★ Warton Crag, Lancashire

How to find

■ **Timing** Adults fly from mid-June to early August, with the season starting and finishing later further north. In the heat of the day they are extremely active and can be very difficult to approach.

■ **Habitat** Both species occur on open, flowery, hilly grassland, often in quite exposed surroundings, but also on woodland edges and glades. The foodplants are various species of violets.

■ **Search tips** Look for the large fritillaries on sunny, not too windy days, but be prepared for views to be fleeting if the weather is very warm – these butterflies can and do vanish in seconds after brief feeding stops. If you search later in the afternoon you may find they spend longer feeding and are more approachable. Explore woodland edges for a chance of finding females egg-laying on violets.

WATCHING TIPS

A large fritillary travelling at full speed is an impressive sight, powering past on its brilliant orange wings. They are attracted to flowers like knapweed and scabious, and if you find a large clump of such flowers you could have the chance to watch the butterflies at length, especially in the morning and evening. High Brown Fritillary sightings should be logged with local butterfly recorders or nature reserve managers – ideally take photos of the underside to confirm identification.

Silver-washed Fritillary

Argynnis paphia

Wingspan 7.5cm

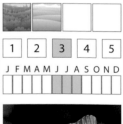

1	2	3	4	5

J	F	M	A	M	J	J	A	S	O	N	D

This is our largest fritillary. Compared to Dark Green and High Brown, it has longer, almost hooked forewings that lack a narrow white border on the upperside, and the underside has a diffuse green/silver wash rather than distinct white spots. The male's upperside forewing has dark streaks along the veins, while the larger female has more typical fritillary spots. A few females are of a very distinctive and beautiful dusky grey-green form, known as 'valesina'. This butterfly is quite common in suitable sunny lowland woods in southern and central England and Wales, with isolated outposts further north, and is patchily widespread in Ireland.

The beautiful, diffuse silvery-grey underwing colouration and more curved forewing shape distinguish this handsome butterfly from our other two large fritillary species. It is also more of a woodland specialist, while Dark Green and High Brown prefer open habitats.

How to find

■ **Timing** Adults emerge in late June, and fly through to mid or late August. They are more approachable in mornings and early evenings, but the stunning courtship chases are most often seen towards the middle of the day.

■ **Habitat** This butterfly may be found in any sunny, flowery woodland, deciduous or coniferous, where the larval foodplant (Common Dog-violet) grows. This can include rural gardens and parklands.

■ **Search tips** Silver-washed Fritillaries are powerful fliers and spend some of their time in the tree canopy, but overall they tend to be more approachable than Dark Green or High Brown Fritillaries, and will feed at length on bramble flowers. Waiting by large bramble clumps in sunny, sheltered places should produce good views. Walk along wide, sunny rides or woodland edges on warm days in the late morning or early afternoon for a chance of seeing courtship flights.

WATCHING TIPS

The courtship chase of the Silver-washed Fritillary is a wonder to behold, the male looping the loop around the female as she flies along. As with the White Admiral, which flies at the same time and in the same places, Silver-washed Fritillaries quickly lose their fresh beauty and become ragged and faded (a consequence of the time they spend climbing over brambles), so seek them out in the second half of June to enjoy them at their magnificent best.

Euphydryas aurinia

Marsh Fritillary

Wingspan 4cm

| 1 | 2 | 3 | **4** | 5 |

J F M A M J J A S O N D

Super sites

★ Strawberry Banks, Gloucestershire
★ The Lizard, Cornwall
★ Aberbargoed Grasslands, Gwent
★ Hod Hill, Dorset
★ Finglandrigg Wood, Cumbria
★ RSPB South Stack, Anglesey

This distinctive, pretty little butterfly is a scarce and vulnerable species, because of its extremely sedentary habits, meaning it is very reluctant to disperse to new or restored habitats. Its basic upperside colour is orange, with creamy and darker brown chequering. The underside is paler, with white chequering and a row of small dark spots on the hindwing border. The caterpillars are highly gregarious, living in conspicuous feeding webs when small. Larger caterpillars become black and very spiny, and leave their webs in groups to bask on sunny days. It is found in south-west England, and sparsely in north-west and south-west Wales, the far west of central Scotland, and patchily throughout Ireland.

How to find

■ **Timing** Adults fly from the end of April until mid-June (a little later in Scotland). The caterpillars' feeding webs can be seen through late summer, and more mature black caterpillars are highly visible in early spring.
■ **Habitat** Despite its name, the Marsh Fritillary will inhabit dry heaths, chalky downland and moors as well as wet grassland. It requires open ground with prolonged sunshine, and plenty of the larval foodplant (Devil's-bit Scabious).
■ **Search tips** Once you have located a colony, these butterflies are usually quite obvious and will even fly and feed in overcast weather – look out for them visiting nectar-rich flowers like knapweeds and thistles. The caterpillars are also very conspicuous in early spring.

WATCHING TIPS

Male Marsh Fritillaries patrol their small territories and intercept passing females – when copulation occurs, it is prolonged, the pair often resting in full view. They also feed frequently, often holding the wings open but with forewings overlying the hindwings, like a moth. Females in search of plants on which to lay eggs are particularly noticeable with their slow, rather laboured low-level flight. They lay large clusters of yellow eggs on the underside of foodplant leaves. Any sightings of Marsh Fritillaries away from known colonies should be passed on to local recorders.

Heath and Glanville Fritillaries

Melitaea athalia, M. cinxia

Wingspan 4.5cm

Heath Fritillary

1	2	3	4	5

J F M A M J J A S O N D

Glanville Fritillary

The Heath and Glanville Fritillaries are close relatives and similar in appearance. They are small with a brown and orange-gold chequered (rather than spotted) upperside. They are best distinguished by the underside pattern – in Glanville this shows black spots within the brown on the hindwing. In Britain, the two were formerly widespread in England but are now scarce and entirely geographically separated. Glanvilles occur on the southern coast of the Isle of Wight – they are also established at a site in Surrey, following introductions. Heath Fritillaries occur sparsely in south-west England, and in north-east Kent and south Essex – some colonies are the result of reintroduction.

Super sites

- ★ RSPB Blean Woods, Kent (Heath Fritillary)
- ★ Haddon Hill, Devon (Heath Fritillary)
- ★ Hutchinson's Bank, Surrey (Glanville Fritillary)
- ★ Bonchurch Down, Isle of Wight (Glanville Fritillary)

How to find

■ **Timing** Both species emerge in late May, and fly through June into early July.

■ **Habitat** Glanville Fritillaries prefer open, sparsely vegetated and warm terrain – on the Isle of Wight this mainly means south-facing undercliff areas. Heath Fritillaries will use open dry habitats too but also lusher sunny woodlands with clearings and glades. Both species use ribwort plantain as the larval foodplant in open conditions, but in woodland Heath Fritillaries favour Common Cow-wheat or foxgloves.

■ **Search tips** These butterflies are rather sluggish and slow fliers, and can usually be found quite easily in the right habitat, feeding at length on flowers such as thistles, vetches or Ox-eye Daisies. Look for them on still, warm and sunny days. Courtship activity happens in the warmest part of the day.

WATCHING TIPS

Both of these butterflies live in rather dense and sedentary colonies, so once you are in the right spot on a good-weather day you should enjoy plenty of good close views of them as they go about their business. As with many other butterflies, courtship and egg-laying takes place after they have warmed up and done some feeding at the start of the day. They are usually approachable and not easily startled. Sightings away from known colonies should be reported to local butterfly recorders.

Papilio machaon

Swallowtail

Wingspan 8.5cm

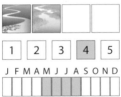

| 1 | 2 | 3 | 4 | 5 |

J F M A M J J A S O N D

A stunning butterfly, in Britain this species is represented by the subspecies *britannicus*. It is confined to the Norfolk Broads, although the less heavily marked continental subspecies *gorganus* occasionally turns up on south and east coasts. It is unmistakeable, with pale yellow wings marked with black veins and borders, and long 'tails' on the hind-wings; the body is also pale yellow with some dark markings. It is a powerful flier. The mature caterpillar is also conspicuous – plump and light green with black stripes, and a forked red organ on its head (the osmeterium) which produces a predator-repelling smell.

There is a wealth of wildlife to see in Broadland in summer, and this beautiful, impressively fast-flying butterfly is the jewel in the crown.

Super sites

★ Hickling Broad, Norfolk
★ RSPB Strumpshaw Fen
★ How Hill, Norfolk

How to find

■ **Timing** Adult Swallowtails emerge in the second half of May, and fly into early July. There is sometimes a small second generation in August. The insects are particularly active in the heat of the day and can be difficult to watch.

■ **Habitat** It is found on sedge- or reed-dominated marshes and fens, where the larval foodplant (milk parsley) grows in abundance – it may also visit flowery gardens close to such marshlands. Migrant *gorganus* individuals can turn up in any open flower-rich habitats.

■ **Search tips** Because its habitat is so open, the Swallowtail is quite easy to spot. As you walk along fenland paths, scan over the tops of the reeds with binoculars and you should eventually see one in its characteristic strong, fast flight. Also check stands of flowers; the species is said to be particularly attracted to deep pink flowers such as ragged robin.

WATCHING TIPS

The Swallowtail is quite easy to see but rather difficult to watch. The damp nature of its habitat means that paths tend to be narrow (often raised banks or boardwalks) and it is not possible to deviate from them. This means that even if the butterfly settles the views may well be distant. For longer views, it is worth waiting by decent clumps of flowers, including milk parsley which will attract egg-laying females. Sightings away from the Broads should be passed on to the local butterfly recorder.

Large, Small and Green-veined Whites

Pieris brassicae,
P. rapae, P. napi

Large White (female)

Wingspan Large W 5cm
Small and Green W 4.5cm

1	2	3	4	5

J F M A M J J A S O N D

Green-veined White (male)

Small White

These three common white butterflies are often lumped together as 'cabbage whites', although only the Large and Small lay their eggs on garden cabbages. All overwinter as pupae, so are among the first butterflies to appear in spring, just behind those that winter in their adult form. The Large White has black at the forewing tips that extends down the outside edge of the wing, while the Small White's wingtip markings are greyer and confined to the tip only. The Green-veined shows prominent dark scaling along the veins on upperside and underside, and the dark wingtip marking is broken into notches. All are abundant in Britain and Ireland, though a little scarcer in upland parts of Scotland.

How to find

■ **Timing** These butterflies fly from late March or early April, and produce one or two more generations through spring and summer; they can be seen well into autumn. In some years, large influxes of migrant Large Whites join resident populations.

■ **Habitat** The Large White is highly nomadic, the Green-veined sedentary, and the Small White between the two. All can occur in almost any kind of well-vegetated sunny habitats, including gardens. The larval foodplants are a variety of domestic and wild brassicas.

■ **Search tips** On a sunny warm day, any country walk (or even glance out of the window) should produce sightings of at least one of these species. For a definite identification, you'll usually need to see the butterfly settled, although the Large White's size and strong flight does stand out. The Green-veined White is more common in rural, well-vegetated habitats such as woodland edges, damp meadows and along hedgerows.

WATCHING TIPS

These butterflies are restless, especially in the heat of the day. Males will fly down to investigate any small white object they notice, in the hope it will be a female. Females that have already mated tilt their abdomens up at the approach of a male, indicating their unwillingness to mate. Planting cabbages and nasturtiums will encourage Large and Small Whites to visit the garden and lay eggs, and you can then observe the gregarious, colourful Large White caterpillars and solitary, well-camouflaged Small White caterpillars as they develop.

Leptidea sinapis, L. juvernica

Wood and Cryptic Wood Whites

Wood White

Wingspan 4.2cm

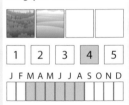

| 1 | 2 | 3 | **4** | 5 |

J F M A M J J A S O N D

Cryptic Wood White

Super sites

★ Monk Wood, Worcs. (Wood White)
★ Powerstock Common, Dorset (Wood White)
★ Oaken Wood, Surrey (Wood White)
★ Peatlands Park ASSI, Co. Armagh (Cryptic Wood White)
★ Raven NNR, Co. Wexford (Cryptic Wood White)

Smaller than the more common white butterflies and much more delicate, these are ethereal beauties. The Wood White has a feeble-looking flight on its very rounded wings, but can fly for long spells without pausing. It rests with wings closed, revealing a soft pale grey wash on the undersides of the wings – the upperside forewing has a smudgy grey tip. The Wood White is sparsely distributed in England, mainly the South-West and Midlands, and is also found in the Burren region of County Clare and County Galway in Ireland. The wood whites widespread elsewhere in Ireland belong to a different species – the virtually identical Cryptic Wood White, which is not present in Britain.

How to find

■ **Timing** Adults fly from early May to late June. In southern England there is often a second generation of Wood Whites in late July/early August, which in some years may be larger than the spring brood.

■ **Habitat** They prefer sheltered, damp, lush places, and are mainly found in woodland glades, edges and rides. However, the Cryptic Wood White will use more open ground, as will the Wood White in the further south-western parts of its range. The larval foodplant is Birdsfoot Trefoil and related species.

■ **Search tips** Look for these butterflies on warm, still days – they are inconspicuous at rest so more easily found towards the middle of the day when activity is highest, particularly males seeking females. The weak-looking, fluttering flight, small size and slightly grey-white look in flight makes them distinguishable from other whites.

WATCHING TIPS

The wood whites have a charming courtship display in which the pair perch facing one another, quickly opening and closing their wings and touching antennae. Once mated and ready to lay eggs, the female has an even more hesitant flight as she searches for foodplants. They can be frustrating to watch as they rarely settle (despite looking barely able to stay airborne) but they will pause to take nectar, more frequently in the morning and evening.

Brimstone

Gonepteryx rhamni

Wingspan 6.5cm

1	2	3	4	5

J F M A M J J A S O N D

The original 'butter fly', the bright yellow male Brimstone is a conspicuous and welcome harbinger of spring. This is our only butterfly other than the colourful 'aristocrats' that overwinter in its adult form. Females are pale green-white and can be mistaken for other whites in flight, but at rest both sexes show the unique wing shape with pointed tips and pronounced venation on all four wings – they also lack black markings, just having a small red spot on each wing. Brimstones are common and widespread in wooded and scrubby habitats, and may visit larger gardens. They range throughout England and Wales, petering out in southern Scotland, and are widespread in Ireland.

Getting close to a Brimstone is much easier in summer than in spring. Any area of sunny open grassland with thistles, knapweeds and other nectar-rich flowers should attract them. In spring, though, a fast-moving yellow blur is often all you'll see.

How to find

■ **Timing** Brimstones emerge from hibernation in March or April, occasionally February. Spring butterflies fly fast and widely in a search for mates, and breed through spring and early summer. There is then a short interval mid-summer, as the spring butterflies have died and the summer generation is yet to emerge. Summer Brimstones fly from late July, and are easier to watch at this time as their focus is on feeding rather than courtship.

■ **Habitat** These butterflies prefer woodland clearings and edges, and more open places with flowers and bushes. The larval foodplants are buckthorn and alder buckthorn. The caterpillars are well camouflaged, resting along a leaf midrib, but their feeding damage can be quite noticeable.

■ **Search tips** A country walk on mild early spring days should produce some sightings of Brimstones, flying strongly along paths and hedgerows. They feed infrequently so views may be brief. In summer, watch out for them visiting stands of thistles, brambles and other nectar-rich flowers in both open and more sheltered spots.

WATCHING TIPS

The bright yellow male is distinctive but you will need a good look to tell the female from other whites – the lack of any black on her wings is usually apparent in flight with good views. In spring, mating pursuits can result in wild spiralling chases high into the sky, sometimes two or more males chasing the same female. Summer butterflies will spend long spells at flowers and are then easy to study at close range. Caterpillars are well-camouflaged but can be found with careful searching of the foodplant in mid-summer.

Anthocharis cardamines

Orange-tip

Wingspan 4.6cm

| 1 | 2 | 3 | 4 | 5 |

J F M A M J J A S O N D

This species winters as a pupa and so appears quite early in the year. However, it only has one generation a year, so you must make the most of it for the few spring weeks it is on the wing. The male is unmistakeable with his orange-tipped forewings. The female could be confused with other whites, but her mottled mossy-green underside is discernible in flight, and her upperside forewing pattern is different, with a broad grey wingtip and a vertical grey marking below it. Orange-tips are found throughout Britain and Ireland, though are scarce in the Scottish Highlands.

The marbled underside of the Orange-tip is the same in both sexes and immediately distinguishes the species from the other 'whites'. On warm spring days they rarely settle but you may have more success in overcast weather.

How to find

■ **Timing** The first Orange-tips may appear in March, but April is the peak month for emergence, and they can be seen into early June. They fly on warm sunny days.

■ **Habitat** These butterflies prefer lush, damp grassland within or alongside woodlands, hedgerows or scrub, and may visit large gardens with plenty of wild flowers. The most often used larval foodplant is Cuckoo-flower, though related species are also used.

■ **Search tips** Male Orange-tips are active and restless, and easy to spot as they roam in search of females, inspecting any small white object along their flight path. On days with intermittent cloud or breezes, they will often settle on Cow Parsley when the air cools, where their mottled green underside affords remarkably good camouflage. The eggs are easy to find on the flower buds of Cuckoo-flower – one tall, skittle-shaped orange egg per plant, as the caterpillars are cannibalistic.

WATCHING TIPS

Orange-tips are surprisingly fast despite a rather weak-looking, fluttering flight. Be prepared to follow one up and down a lane for some time before it settles and allows closer views. If you find a good clump of Cuckoo-flowers, watch out for visiting females, settling on the flower heads to carefully lay a single egg on a flower bud – both sexes will also visit Cuckoo-flowers for nectar. As with other whites, females only mate once and after that will reject amorous males by holding their abdomens aloft. This species has spread north in recent years – sightings on the edge of its range will be of interest to your local butterfly recorder.

Clouded Yellow

Colias croceus

Wingspan 5.6cm

1	2	3	4	5

J F M A M J J A S O N D

If you manage to photograph a Clouded Yellow in flight, the upperside forewing edge will reveal its sex – solid black in males, black with yellow spots in females. This wing pattern sometimes shows through on resting individuals too if they are strongly backlit.

Second to the Painted Lady, this is our most common migrant butterfly, occurring only as a visitor as it usually cannot survive the winter here in any of its life stages. Numbers reaching our shores vary greatly from year to year, determined by how prolific it is in its resident range in north Africa and southern Europe. It flies strongly and may be seen anywhere in Britain, although sightings are most numerous along southern and eastern coasts. Its deep orange-yellow colour and extensive black wing edges make it quite distinctive in flight, although the pale *helice* form of the female may be confused with various whites.

How to find

■ **Timing** The first migrants arrive in May, and there are further arrivals through summer and into autumn. Southerly and south-easterly winds increase the chances of large numbers making landfall.

■ **Habitat** Clouded Yellows are most often encountered in open, rather windswept countryside, ranging from marshland to moorland, farmland to open meadows. They seek out clovers and related species for egg-laying.

■ **Search tips** This butterfly is highly mobile and unpredictable in its appearances, and in many years even a keen wildlife-watcher may encounter few or none, simply from not being in the right place at the right time. Coastal areas in mid-summer offer the best chance. Use binoculars to scan at all distances. Clouded Yellows do pause to feed and may spend some time at a single flower before powering away again.

WATCHING TIPS

Witnessing a mass arrival of Clouded Yellows is a rare and unforgettable experience. Keep an eye on wildlife-watching social media, such as Butterfly Conservation's Migrant Watch page, through spring and if early signs are good for a 'big year', aim to plan a trip to a flowery coastal headland when the wind is blowing the right way. Arrivals can continue well into October. In big years, there is an increased chance of encountering the much rarer Pale Clouded Yellow and Berger's Clouded Yellow, both of which are paler than Clouded Yellow, but remember the pale *helice* form of the Clouded Yellow.

Hamearis lucina

Duke of Burgundy

Wingspan 3.2cm

1	2	3	**4**	5

J F M A M J J A S O N D

This much-coveted species is often known simply as 'The Duke'.

This little butterfly is the sole British representative of the mainly tropical family Riodinidae, though older books may refer to it as 'Duke of Burgundy Fritillary', reflecting similarities in appearance with fritillaries. It has brown and orange-brown chequered uppersides to its wings, and the brown underside hindwing is marked with a double row of large white spots. The Duke of Burgundy is an uncommon and declining species. It is mainly restricted to central southern counties in England, but has a few outposts in north-west and north-east England.

Super sites

★ Bonsai Bank, Kent
★ Noar Hill, Hampshire
★ Ivinghoe Beacon, Hertfordshire
★ Gait Barrows, Lancashire
★ Arnside Knott, Lancashire

How to find

■ **Timing** Adults emerge in late April and fly until the end of May or early June (later in the northerly colonies). Occasionally southern colonies produce a small second brood.

■ **Habitat** It is mainly found on open chalky grassland with plenty of flowers and some scrub, and a small number of colonies survive in coppiced woodland but fewer now than historically. Its larval foodplants are Cowslip and Primrose.

■ **Search tips** These butterflies have a short but intense flight season. Look for them on warm sunny days. Explore the habitat slowly as they are inconspicuous when not flying, and rarely visit flowers for nectar. From mid-morning, males can often be seen resting quite high on hawthorn leaves, waiting to intercept passing females.

WATCHING TIPS

The territorial males will return to their favourite perch after each darting chase after another insect, and will usually tolerate close inspection. Females in search of egg-laying sites fly more slowly and at low levels, pausing briefly to lay one or a few eggs on the underside of particularly lush green leaves of the foodplant. A close look at individuals of both sexes reveals the interesting quirk that, although they look very similar in colours and pattern, females have six full-sized legs, males just four. Duke of Burgundy sightings away from known colonies should be reported to the local butterfly recorder.

Large, Small and Essex Skippers

Ochlodes sylvanus,
Thymelicus sylvestris, T. lineola

Wingspan Large S 3.3cm
Small and Essex S 3cm

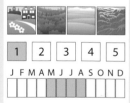

1	2	3	4	5

J F M A M J J A S O N D

Large Skipper

Small Skipper

These three 'golden' skipper butterflies are all common and widespread in England, with Small and Large also common in Wales. They have orange-yellow wings, with a dark line (scent brand) on the forewing in the male, and rest in a characteristic posture with the forewings raised and hindwings flat. They can be challenging to identify. Large Skipper has pale and dark chequering on its wings. The plainer Small and Essex are best told by the colour of their antenna tip undersides – black in Essex, brown in Small. In a few regions confusion is possible with the rare Lulworth and Silver-spotted Skippers – (see p109 and 110).

Essex Skipper

How to find

■ **Timing** Large Skippers emerge in late May and fly into July, while the other two appear in June and fly into mid-August, the Essex typically first appearing about a week after the Small.

■ **Habitat** All three are dependent on long grass. The Large prefers sheltered grassy areas with some scrub, such as woodland edges and glades, while the Small and Essex are found in more open habitats – all three may occur in town parks and larger gardens. Their larval foodplants are cocksfoot and various other grasses.

■ **Search tips** These skippers have a buzzing, moth-like flight that can be hard to follow, especially with the Small and Essex which fly through rather than above the grass sward, weaving around the stems. When resting they are eye-catching with bright golden wings and often choose raised perches on which to bask. Watch for them visiting flowers to feed – the Large particularly likes bramble, while the others enjoy knapweeds and thistles.

WATCHING TIPS

These skippers are rather unassuming but delightfully lively and approachable, with the Small and Essex sometimes occurring in profusion. Males are territorial and will chase others from their small patch, while females will happily share a flower. Egg-laying females select rolled-up grass blades and push their abdomen tips against the sheath, laying a few eggs at a time. The Essex Skipper is spreading its range north and west, and is probably the species most likely to turn up in smaller gardens. Sightings beyond its current range should be passed on to local butterfly recorders.

Thymelicus acteon

Lulworth Skipper

Wingspan 2.6cm

1	2	3	**4**	5

J F M A M J J A S O N D

This skipper resembles the Small and Essex Skippers, but is even smaller than them and has a somewhat duskier appearance, with a vague paler golden half-circle marking on the upperside forewing in the female. It is extremely localised, occurring only in a small area of south Dorset between Bournemouth and Weymouth, but within its range it can be very numerous. It has never been recorded far from this core range, but is widespread on the continent, indicating that Britain represents the northerly limits of its distribution – climate change may allow it to spread northwards. Small and Essex Skippers may occur alongside it, so care must be taken with identification.

Skippers are approachable when roosting at dusk and dawn, allowing close examination.

Super sites

★ Durlston CP (Dorset)
★ Corfe Castle (Dorset)
★ Ballard Down (Dorset)

How to find

■ **Timing** Adults emerge in early June, occasionally earlier, and are on the wing until early August.
■ **Habitat** It requires warm (so usually south-facing) grassy slopes, usually on chalky or limestone soil as this supports good growth of the larval foodplant, tor-grass.
■ **Search tips** Finding the Lulworth Skipper involves slow exploration of its habitat on a warm, not too windy day, and careful checking of all small 'golden' skippers you encounter. Like Small and Essex Skippers, it often flies through the grass sward and is not easy to track in flight, but often stops for nectar or to bask with wings partly open. At the right place and time, Lulworth Skippers will greatly outnumber other skipper species.

WATCHING TIPS

These little butterflies seem to lose their fresh appearance very quickly and become dark and drab, so it's worth trying to catch them near the start of their season. As evening draws in, you may see dozens resting together at the tops of grass seed heads with wings open, soaking up the last of the day's sunshine. The following morning many will still be in their previous positions and as soon as the ground is sunlit they will open their wings again to bask before beginning their day's work. Notify the local butterfly recorder of any Lulworth Skipper sightings away from usual sites.

Silver-spotted Skipper

Hesperia comma

Wingspan 3.4cm

1	2	3	4	5

J F M A M J J A S O N D

This attractive skipper is a late-summer species.

This downland specialist is the most distinctly marked of Britain's five 'golden' skippers. It is the size of Large Skipper and, like it, has a chequered wing pattern, but its pale markings are much whiter and more prominent, especially on the underside hindwing. The male also has a rather indistinct black sex-brand on the upperside forewing. It is a late-season flyer and by the time it emerges there will be very few Large Skippers still on the wing. The Silver-spotted Skipper is confined to the North and South Downs, the Chilterns and similar habitats in the South-West, and is gradually increasing in number, though it was formerly much more widespread.

Super sites

* ★ Lullingstone Heath, East Sussex
* ★ Malling Down, East Sussex
* ★ Lydden Down, Kent
* ★ Old Winchester Hill, Hampshire
* ★ Aston Rowant, Oxfordshire

How to find

■ **Timing** Adult Silver-spotted Skippers fly from late July through August and the first half of September.

■ **Habitat** They prefer a habitat combining well-grazed grassy downland with patches of longer grass and plenty of flowers, on warm, often south-facing slopes. The larval foodplant is Sheep's Fescue, a type of grass.

■ **Search tips** Search for these skippers on warm days without too strong a breeze. Like other 'golden' skippers, they are avid nectar-feeders, and if you station yourself near a good stand of knapweeds you are likely to have sightings before long – they also visit much smaller flowers in well-grazed grass. There are unlikely to be many other skippers of any species on the wing by the time the Silver-spotteds are out in force.

WATCHING TIPS

Silver-spotted Skippers are fast on the wing and hard to follow, so patience is needed. They will pause often, to bask on warm bare ground as well as to take nectar, and are usually easily approachable as long as you move slowly. Males will rest on more prominent perches when searching for a mate, and if they manage to intercept a passing female that has not yet mated, the pair copulate on the ground after a short chase. Males may also pursue other small butterflies, so could direct your attention to an otherwise inconspicuous female Adonis Blue, or other interesting species.

Dingy and Grizzled Skippers

Dingy Skipper

Wingspan Dingy S 3.2cm
Grizzled S 2.8cm

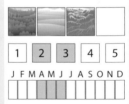

1	2	3	4	5

J F M A M J J A S O N D

Grizzled Skipper

Although not closely related, these two skippers have a similar flight season and habitat preferences. They are also our only skippers that are grey rather than brown or golden. Both are often mistaken for moths but have a broad head with large, well-spaced eyes and slightly hook-tipped antennae. The Dingy Skipper has a subtle grey pattern, quite beautiful when fresh, while the Grizzled is dark grey with small white spots – both are paler on the underside. They hold their wings spread flat; on cooler days Dingy often holds the forewings swept back, like a moth. The Grizzled is found in south and central England and parts of Wales, while the Dingy is more widespread and occurs patchily further north to Scotland, and in Ireland.

Their relatively small, drab-coloured wings and their stocky big bodies lead to confusion with moths. However, both of these species have the distinctive skipper head shape with big, well-spaced eyes and hook-tipped antennae.

How to find

■ **Timing** These two species are spring flyers, appearing in April or early May and flying into mid-June. Grizzled is usually out a little ahead of Dingy.

■ **Habitat** Both species like warm, open habitats with patchy ground vegetation and plenty of flowers for nectar. They may be found in woodland clearings, on well-grazed downland, on heath edges and on waste ground. The Dingy Skipper's larval foodplants are bird's foot trefoil and other vetches, while the Grizzled uses various plants including Wild Strawberry, Agrimony and Creeping Cinquefoil.

■ **Search tips** Both species have a fast, erratic, low-level flight that is very hard to follow, particularly in the case of the tiny Grizzled Skipper. They also share their habitat with various grey day-flying moths. Obtaining good views and thus identification is therefore a matter of patience. You will have most success on a still, sunny day – visit in the morning when the butterflies are still basking.

WATCHING TIPS

These skippers do not feed as avidly as the 'golden' species but do need to bask at length in the cooler parts of the day, often posing photogenically atop a stone or dead flower-head. Males are territorial and chase other similar-sized insects. Females with eggs to lay fly a little slower than usual and so may be followed more easily as they seek out suitable plants. If visiting a site that hosts both species, go in the first half of May, when there should still be Grizzled Skippers flying, and the Dingies will be looking their best.

Chequered Skipper

Carterocephalus palaemon

Wingspan 3cm

| 1 | 2 | 3 | 4 | 5 |

J F M A M J J A S O N D

The Chequered Skipper provides a cautionary tale, having become extinct in England in 1976 following habitat loss in the Midlands woodlands where it lived. It still survives in a small area of north-west Scotland, where numbers are stable within its limited range. It is a dark brown skipper with prominent large creamy spots on all wing uppersides – the underside is much paler. It is more similar to Dingy and Grizzled Skippers than to the 'golden' skippers in its general appearance, resting with wings spread flat rather than the curious angled position used by the 'golden' species.

This rare butterfly may soon be back in eastern England – conservation bodies are working on a reintroduction project.

Super sites

★ RSPB Glenborrodale
★ Ariundle Wood, Scottish Highlands

How to find

■ **Timing** Adults appear in the second half of May, and fly until mid-June.

■ **Habitat** It prefers warm, sheltered grassland on the edge of woodlands or in clearings or glades, often close to water, where its larval foodplant (purple moor-grass) grows well.

■ **Search tips** Chequered Skippers tend not to form dense colonies, so can be difficult to find even in the right habitat. Allow plenty of time, especially with the changeable weather that prevails in north-west Scotland. When searching, pay attention to tall shrubs on the edge of open ground, as males use such vantage points to survey their territory, and watch out for them visiting ground ivy and other purple- and blue-flowered plants for nectar.

WATCHING TIPS

These are typical charming skippers, with fast flight but regular pauses to feed and to bask in the sunshine, on foliage, bare ground or a warm rock. However, when the sun goes in they quickly disappear into the shelter of the deep grass. Males are the more active sex, constantly alert to the possibility of intercepting a female. Courtship is a hasty tumble down to ground level, where the pair mates. The female lays her eggs singly on the centre of a grass blade, rather than hidden in a furled leaf as with the 'golden' skippers – the round, pale eggs can be quite conspicuous.

Callophrys rubi

Green Hairstreak

Wingspan 3cm

1	2	3	4	5

J F M A M J J A S O N D

This delightful little early-season butterfly is our only species with truly bright green wings, and is unmistakeable if seen well. The apparent green colour, which affords very good camouflage, is produced by refraction of light from the modified wing scales, rather than actual green pigment. The wings are always closed at rest, showing their shining green underside with a variable white 'hairstreak' on the hindwing (often reduced to just a white dot or two). The uppersides are plain brown, and in flight the butterfly could be confused with a blue species at first glance. It is extremely widespread throughout Britain and Ireland, but its distribution is patchy, avoiding built-up and heavily wooded areas.

The Green Hairstreak is inconspicuous at rest but spend some time scanning sunlit shrubs – basking butterflies will flutter to a new, even more sunny spot every so often.

How to find

■ **Timing** Adults emerge in mid- or late April and fly into mid-June, by which time they are often very faded and show little of their characteristic green sheen.

■ **Habitat** It is found in warm areas where plenty of scrub grows, including woodland edges, hedgerows, waste ground, heath and moorland and chalky grassland. The larval foodplants include Birdsfoot Trefoil, broom and gorse, but many other shrubby and herbaceous species as well.

■ **Search tips** Even the sharpest eyes can have trouble spotting this small green butterfly resting on a large green shrub, and numbers are low at many sites. Search patiently at spots where there are tall bushes adjacent to patches of flowers and look out for any small, drab-looking butterfly in flight – track it carefully to examine it when it lands.

WATCHING TIPS

Green Hairstreaks are inconspicuous but usually quite relaxed about being studied closely. They will perch at length on a high, sunlit leaf, leaning sideways to absorb as much of the sun's warmth as they can. They do not take nectar as avidly as many spring butterflies but females in particular will come down to feed on trefoils and other low-growing flowers, in between searching out foodplants. This butterfly is declining, but on the other hand is quite effective at dispersing to new sites, so look out for it in any suitable habitat and pass on observations to your local butterfly recorder.

Brown Hairstreak

Thecla betulae

Wingspan 4cm

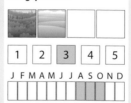

1	2	3	4	5

J F M A M J J A S O N D

Its rich colours befit this most autumnal of butterflies.

This relatively large hairstreak has the latest flight season of any single-generation British butterfly. It rests with wings open or closed – when open it reveals dark velvety brown uppersides with a broad orange band on the forewing in the female (smaller in the male). In flight, it can be confused with the similarly coloured Gatekeeper. The underside is light orange-brown with a double white 'hairstreak' on the hindwing, and one dark and one light streak on the forewing. The lower body and legs are strikingly white. Brown Hairstreaks are rather uncommon and localised. Their strongholds are south and south-west England, a few parts of the Midlands, south-west Wales, and the Burren area of west Ireland.

Super sites

- ★ RSPB Pulborough Brooks, West Sussex
- ★ RSPB Otmoor, Oxfordshire
- ★ Noar Hill, Hampshire
- ★ Orley Common, Devon
- ★ Welsh Wildlife Centre, Pembrokeshire

How to find

■ **Timing** Adults emerge at the end of July and fly through August and into September. They become easier to see later in the season, when they spend less time in the high treetops.

■ **Habitat** This butterfly prefers warm areas with scrub (in particular Blackthorn, the larval foodplant), and some tall trees (ideally Ash), around which courtship activity is centred.

■ **Search tips** Scanning the tops of Ash trees with binoculars may produce sightings of territory-holding males and visiting females. As summer progresses the females are more likely to be seen lower down, basking on leaves and perhaps visiting flowers such as brambles. Choose a warm or hot day and concentrate mainly on eye-level and above when you are scanning the vegetation for activity.

WATCHING TIPS

Like some other hairstreaks, this species feeds on aphid honeydew, and can therefore spend much time high in the trees. Males occupy the tops of 'master' Ash trees and rarely descend to an easily viewable height, but you can see them basking and having territorial squabbles through binoculars. Females visit flowers more often and come lower down to lay their eggs, seeking out fresh growth on Blackthorn bushes. They lay their distinctive and noticeable disc-shaped white eggs at a twig fork in a shaded spot.

Satyrium pruni

Black Hairstreak

Wingspan 3.7cm

| 1 | 2 | 3 | 4 | 5 |

J F M A M J J A S O N D

This rather unobtrusive hairstreak is one of Britain's rarest butterflies, though usually not too difficult to see if you visit one of its few sites at the right time and are willing to be patient. It always rests with wings closed, showing a light brown underside with an orange border, broadening towards the bottom of the hindwing and edged with black dots, and a white 'hairstreak' on both forewing and hindwing, following a smooth curve unlike the jagged 'W' shape on the otherwise very similar White-letter Hairstreak. The hindwing bears a short black, white-edged 'tail'. The uppersides are plain mid-brown. It is confined to a few sites in a small area of the east Midlands.

This is a very vulnerable, rare butterfly, unlikely to be able to disperse to new sites without some human intervention. Most of its colonies are under close protection by conservation organisations.

How to find

■ **Timing** Adults emerge in early June, and their short flight season is virtually over by the start of July.

■ **Habitat** It is found in warm, sheltered open woodland, woodland edges and sometimes mature hedgerows, where the foodplant (Blackthorn) occurs.

■ **Search tips** Black Hairstreaks are avid feeders on aphid honeydew, which means they spend much time in high treetops, but they are also attracted down to feed on brambles, privet and other flowers. Use binoculars to look for them in the trees, or even a telescope, and be prepared to 'stake out' large clumps of flowering brambles. The Wildlife Trust for Bedfordshire, Cambridgeshire and Northamptonshire, which manages the key site Glapthorn Cow Pasture, holds walks in summer to look for the species.

WATCHING TIPS

The spiralling courtship dances of Black Hairstreaks reveal another side to this somewhat sluggish-seeming butterfly, and it is worth watching the treetops to try to see it. Otherwise, you are most likely to enjoy prolonged views when they descend to feed, moving deliberately from flower to flower while moving their hindwings in a curious circling 'grinding' motion, characteristic of hairstreaks and their relatives. This very rare butterfly is notoriously poor at dispersing to new sites, so any sightings away from known colonies should be reported to butterfly recorders.

White-letter Hairstreak

Satyrium w-album

Wingspan 3cm

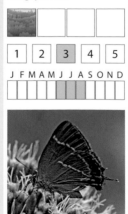

1	2	3	4	5

J F M A M J J A S O N D

This species is very like the Black Hairstreak, differing in its more angular, zigzagging white hindwing 'hairstreak' (forming a W-shape, giving the species its name), and its less extensive orange markings, which fade away halfway up the hindwing. The brown tone of its wings is also a little cooler and greyer. The uppersides, only revealed in flight, are plain light brown. Its dependency on elm trees meant it declined severely after Dutch Elm Disease devastated Britain's mature elms in the 1970s, but it can survive on sucker growth around dead and dying elms too. It is widespread in England and Wales, only really missing from the far south-west and north-west, but rarely abundant.

Anywhere with a decent amount of elm growth could support this widespread butterfly, but it is one of our most difficult species to see. You will need some good luck to encounter one at close enough range to see it well without binoculars.

How to find

■ **Timing** Adults emerge in late June, and fly into mid-August.

■ **Habitat** Any habitat where elms are found can support colonies of White-letter Hairstreaks, though it is easier to observe at sites where elms grow on woodland edges. Its preferred foodplant is Wych Elm but it uses other related species as well.

■ **Search tips** This butterfly spends most of its life high in the trees, where it feeds on aphid honeydew. It is thus a challenging species to find, and you'll probably need binoculars to have a sighting. Scan the outer leaves, looking for the small dancing shape around the canopy. A telescope can be helpful too. The butterflies do sometimes come lower down, especially in the morning and evening, and feed from flowers such as brambles. Colonies based around elm growth in hedgerows may provide easier sightings.

WATCHING TIPS

The White-letter Hairstreak's daily routine is similar to that of other butterflies. Once warmed up enough for flight, it spends time feeding at the start and end of the day, and in the middle of the day courtship activity takes place. This is hard to watch, as is the egg-laying behaviour of the female. She climbs among the elm flower buds, laying single, flattened grey eggs on them. The easiest way to watch this butterfly is when it comes to flowers for nectar but you may need considerable patience.

Favonius quercus

Purple Hairstreak

Wingspan 3.5cm

1	2	3	4	5

J F M A M J J A S O N D

This is the most common of our hairstreaks but by no means the easiest to see. With clear views it is distinctive, showing very dark uppersides with extensive purple gloss on both wings in the male, but just a patch on the forewing in females. The underside is soft grey with a dark-edged white 'hairstreak' on both wings, a duskier line between the 'hairstreak' and the wing edge, and a black spot circled with orange on the hindwing margin, just in front of the tiny white 'tail'. Purple Hairstreaks are widespread in England and Wales, becoming scarcer in the north, and with some colonies in central Scotland and scattered sparsely in Ireland.

Sometimes confused with White-letter Hairstreak, the Purple Hairstreak has a greyer underside, with a bolder, black-edged white hairstreak. It is also rarely seen anywhere other than in an oak tree.

How to find

- **Timing** Adults fly from late June through to the end of August, a little later in Scotland.
- **Habitat** Its dependence on oak trees (preferably Pedunculate but also Sessile, Holm and Turkey oaks) dictates its habitat choices – most Purple Hairstreak colonies are in deciduous woodland. However, just one or two mature oaks within parkland or a garden could also support a colony.
- **Search tips** Binoculars are very useful, if not essential, to find this species – as with most other hairstreaks. Scan the tops of oak trees, and if possible make your way to a high point overlooking the woodland so you can scan the canopy from above. Purple Hairstreaks do occasionally descend to feed from flowers, and your best chance of catching this is to explore the woodland in the morning or evening, checking stands of brambles particularly – they may also drink from puddles. You may even have some success by spraying a sugar solution over lower tree foliage with a water-misting spray bottle.

WATCHING TIPS

This fairly common species is a mystery to most wildlife-watchers because of its arboreal habits, but if you manage to find a good elevated vantage point you can see much of its interesting behaviour up in the trees. This is one butterfly species that may be successfully watched through a telescope! Colonies can hold many individuals and you may see groups engaged in chases over the treetops. Also look out for females basking on leaves in between their egg-laying forays.

Chalkhill and Adonis Blues

Polyommatus coridon, P. bellargus

Chalkhill Blue

Wingspan 3.7cm

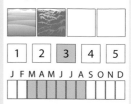

1	2	3	4	5

J F M A M J J A S O N D

Adonis Blue

These two beautiful butterflies are closely related and both depend on warm chalky or limestone downland. The Chalkhill Blue is the more widespread, occurring across suitable habitat in southern and south-east England. The Adonis occurs in broadly the same areas but has fewer colonies. Males of the two species have distinctive uppersides – the Chalkhill is a light, creamy blue with a row of dark spots on the hindwing margins, while the Adonis is brilliant turquoise-blue with a black-and-white chequered wing margin. Females are very similar – dark brown on the upperside with a border of faint orange spots.

Super sites

- ★ Durlston CP, Dorset
- ★ Cerne Giant, Dorset
- ★ Aston Rowant, Oxfordshire
- ★ Martin Down, Hampshire
- ★ Old Winchester Hill, Hampshire
- ★ Castle Hill NNR, West Sussex
- ★ Lydden Down, Kent

How to find

■ **Timing** Chalkhill Blues are on the wing from mid-July to the end of August or just into September. Adonis Blues have two generations a year, the first flying from mid-May to mid-June, and the second on the wing from early August to mid-September.

■ **Habitat** Both species occur on open, grassy and flowery downland with a mainly short turf where their larval foodplant (Horseshoe Vetch) is found, along with ants, which attend the larvae. The Adonis Blue needs warm ground in its larval stage, so occurs on south-facing slopes.

■ **Search tips** These butterflies are easy to see but quick on the wing and usually fly low, so walk slowly through their habitat (keeping to paths or trails to avoid accidentally stepping on mating pairs). If you search in the heat of the day, they will be highly active, but at sunrise and again in the evening they are more approachable and spend longer feeding and basking. Males often take nutrients from around muddy puddles or animal droppings.

WATCHING TIPS

These butterflies have a lot to accomplish in their short flight seasons, and will be very active on warm dry days, feeding, courting, mating and egg-laying. They can be watched easily because of the open grasslands they inhabit, but resist the temptation to pursue them and risk trampling their habitat. Colonies of Adonis Blue are often dense so if one butterfly darts away from view another will soon replace it. Look out for females visiting Horseshoe Vetches to lay their eggs.

Polyommatus icarus

Common Blue

Wingspan 3.3cm

1	2	3	4	5

J F M A M J J A S O N D

Though no longer as common as its name suggests, this is still the most likely 'blue' to be seen in open countryside. The male has violet-blue uppersides, the female is brown with variable amounts of blue towards the centre of her wings and orange markings along the hindwing border. The brownest females can be confused with Brown Argus and Northern Brown Argus – (see p120) for identification tips. The underside is light brown with white-circled black spots and orange border markings – this immediately separates it from the Holly Blue which has a pale blue underside with a few tiny black spots. The Common Blue is found throughout Britain (subspecies *icarus*) and Ireland (the brighter subspecies *mariscolore*).

Male Common Blues have entirely blue uppersides. Females vary from entirely brown to very blue, but even the bluest females always show a row of orange spots along each wing border.

How to find

■ **Timing** Common Blues mainly produce two, occasionally three generations a year, so are on the wing from May through to September or later with a short hiatus between generations in early July. Northern and Irish populations have one generation, flying from June to August or early September.

■ **Habitat** This butterfly occurs in all kinds of open, grassy habitats, including on heath, woodland and moor edges, farmland set-aside, marshland and sometimes parks and larger gardens. The main larval foodplant is Birdsfoot Trefoil, but related vetch species may also be used.

■ **Search tips** The male is striking in flight and at rest, and readily spotted as you walk in good habitat – the female is less active and much less conspicuous but any small, dark butterfly in suitable habitat is likely to be a female of this species. Both sexes frequently visit flowers of all kinds – in particular, check knapweeds, ragwort and thistles.

WATCHING TIPS

Watch Common Blue males patrolling their territories, in search of females but also frequently pausing to feed and bask. Some females are nearly as blue as males, their male-like colouration perhaps protecting them from excessive male attention. The combination of blue and orange on their uppersides makes them one of Britain's prettiest butterflies, well worth a close look. Mating pairs often perch for long periods atop a flower head, giving great views, rather than copulating on the ground as some other blues do.

Brown and Northern Brown Argus

Aricia agestis, A. artaxerxes

Brown Argus

1	2	3	4	5

J F M A M J J A S O N D

Brown Argus

J F M A M J J A S O N D

Northern Brown Argus

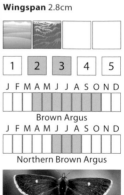

Northern Brown Argus

These small butterflies are part of the 'blues' group but their uppersides are brown. Both can be confused with the Common Blue female but have a different spot arrangement, and more orange and a broader white border on the forewing uppersides. Also, the Northern Brown Argus usually has a white central spot on the forewing upperside, and white spots without black centres on the undersides. Brown Argus occurs in southern and eastern England, parts of Wales and the far South-West; Northern Brown Argus occurs patchily in north-west England (subspecies *salmacis*) and Scotland (the white-spotted subspecies *artaxerxes*). There are Brown Argus populations in north Wales that are of uncertain identity.

Although not colourful, the two brown argus species are attractive and charming with a feisty character that belies their diminutive size. They lack a blue dusting on their brown uppersides, distinguishing them from the female Common Blue.

How to find

■ **Timing** The Brown Argus produces two generations a year, flying in May and June, and from mid-July to mid-September. The Northern Brown Argus has a single generation, flying in June and early July.

■ **Habitat** These butterflies like open, hilly, grassy habitats, with plenty of flowers and warm bare ground, and the larval foodplant, common rock-rose.

■ **Search tips** Very small and not colourful, these butterflies are easily overlooked. As you walk, look for little dark butterflies with a fast flitting flight low over the ground. Follow them carefully as they will soon stop to bask or feed. Like other blues, they will roost perched head-down on the top of a grass head or flower head, and once you 'have your eye in' you can easily find them like this in the evenings, early mornings, or spells of overcast weather on otherwise sunny days.

WATCHING TIPS

Both of the brown argus species are rather enigmatic little butterflies, especially in northern England where their ranges come close to overlapping and they are most similar to each other. They are generally approachable and easy to watch and photograph as they feed and bask – males of both are also aggressive and will chase other butterflies much larger than themselves. The distinctive white-spotted *artaxerxes* is a highlight of any Scottish butterfly-finding trip.

Celastrina argiolus

Holly Blue

Wingspan 3cm

| 1 | 2 | 3 | 4 | 5 |

J F M A M J J A S O N D

Blue butterflies are often presumed to be Common Blues, but in many cases they are more likely to be the showy, wide-ranging Holly Blue, which has quite different habitat needs to all of our other open country-loving blue butterflies. This species is easily identified by its plain silver-blue underside, marked with a sparse pattern of small black dots. Both sexes have blue uppersides, females have an obvious dark border to the forewings (particularly heavy and extensive in the summer generation). The wings are typically held half-open rather than fully spread at rest. The Holly Blue is widespread in England and Wales, and present over most of Ireland as well.

The Holly Blue has the plainest underside of any of our blue butterflies, but makes up for this with a beautiful light violet-blue upperside, bordered with dusky blackish in females.

How to find

■ **Timing** Holly Blues winter as pupae and so are among the first spring butterflies to appear. The spring brood flies from late March or April into June, and the second brood is on the wing from mid-July into September.

■ **Habitat** This butterfly can be seen in all habitats that are reasonably sunny and support its larval foodplants (holly and ivy). It is often encountered in gardens of all sizes, town parks, and along woodland edges, and is one of the species most likely to actually breed in gardens.

■ **Search tips** Look for Holly Blues on warm days from early spring, along hedgerows and around the garden. They often fly at high levels and are very restless, pausing to feed much less often than other blues. In spring, they will mainly seek out holly trees and bushes for egg-laying, while the summer generation lay their eggs primarily on ivy. Males may also come to the ground to drink from puddles.

WATCHING TIPS

This butterfly is one good reason to allow ivy to proliferate in the garden. Watching a summer-generation female exploring a large clump of ivy gives the opportunity for great views as she moves over the plant, placing single white eggs on unopened flower buds. If your garden has both holly and ivy, so much the better, particularly if the holly is female (berry-bearing). Holly Blues do not spend much time visiting nectar sources but do enjoy basking in sunshine.

Silver-studded Blue

Plebejus argus

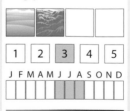

Wingspan 2.9cm

1	2	3	4	5

J F M A M J J A S O N D

The silver spots on the underside are not very conspicuous but sparkle in the light if you look from the right angle.

This is an uncommon and distinctive small blue butterfly. The male has violet-blue uppersides with a dark border and white fringe. The female is dark brown on the upperside. Both sexes have silver-blue spots along the edge of the hindwing underside. It has a very patchy distribution; mainly heathlands in Dorset, Hampshire and Surrey, but there are also significant colonies near the coast in Cornwall, East Anglia and Wales. Colonies can be very dense and give the impression of great abundance, even though the species as a whole is a rarity, having declined through habitat loss. There have been some successful reintroductions, including in Norfolk and Merseyside.

Super sites

★ Thursley Common, Surrey
★ South Stack, Anglesey
★ RSPB Westleton Heath

How to find

■ **Timing** Silver-studded Blues fly from mid-June to early August, with several days separating the appearance of the first males and the first females.

■ **Habitat** This butterfly is associated with warm sunny heathlands inland, coastal sea-cliff heaths, and grassland on limestone or chalky soils. Its larval foodplants include Bell Heather, Cross-leaved Heath, and, in grassland, Birdsfoot Trefoil and Common Rock-rose.

■ **Search tips** If you arrive at a good site near the start of the flight season you are likely to be confronted by many hundreds of male Silver-studded Blues. A few weeks later, locating the surviving butterflies can be very difficult. Look for small, dark, fast-flying butterflies skimming low over the tops of the heather or grassland, or basking with wings open, even when there is cloud cover.

WATCHING TIPS

Visit a colony early in the day (and the season) to enjoy the spectacle of dozens of males basking in close proximity, decorating the heather with their spread violet wings. They take this opportunity to soak up the sun before seeking out newly emerged females. Competition to mate is considerable, and if you see a female at all she will probably already be paired with a male. Later in the season, males are fewer in number and you could see females searching for places to lay their eggs.

Maculinea arion

Large Blue

Wingspan 4.5cm

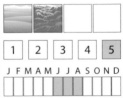

1	2	3	4	5

J F M A M J J A S O N D

The successful reintroduction of Large Blue to Britain is a great triumph for science and conservation.

This butterfly has a long and complex history in Britain, having declined to eventual extinction in the 1970s. That was the end of the British subspecies *eutyphron*, but the subspecies *arion* was then successfully reintroduced from Sweden over the following decades. Returning the species hinged on fully understanding its unique and complex life cycle, and the crucial role played by a species of ant. Large Blues are distinctive, showing mid-blue uppersides with dark borders and some dark spots on the forewing. The undersides are the typical pale beige with white-circled black spots, and barely any orange markings in the hindwing margin.

Super sites

★ Collard Hill, Somerset
★ Green Down, Somerset

How to find

■ **Timing** Large Blues have a short flight season, from mid-June to early July.

■ **Habitat** The sites where it is now established are hilly, warm grasslands with short turf, populations of the ant *Myrmica sabuleti*, and plenty of Wild Thyme (the larval foodplant, up until the time that the part-grown larva is taken by the ants into their nests, when it feeds on ant larvae but compensates the ants by providing them with honeydew).

■ **Search tips** Most of the reintroduction sites are not open to the public, to protect the habitat from damage and the butterfly from collectors (it has full legal protection). You can see it at an open access site at Collard Hill in Somerset, or by joining an organised visit (arranged through the Somerset Wildlife Trust and Butterfly Conservation) to a private site, Green Down. Large Blues visit thistles and Wild Thyme for nectar, and will bask with wings open in overcast weather to warm up.

WATCHING TIPS

To watch this butterfly is to witness conservation in action. Reintroduction is by no means an easy solution to declines and national extinctions – similar schemes for the Large Copper have so far failed. The most intriguing aspect of the butterfly's life – its time spent as a larva and pupa in ants' nests – is sadly impossible to observe as it takes place underground, but you can see females exploring unopened Wild Thyme flower heads, laying single eggs on suitable spots. Even at this stage, the ants take an interest in both egg and adult butterfly.

Small Blue

Cupido minimus

Wingspan 2.4cm

1	2	3	4	5

J F M A M J J A S O N D

The faint white circles around the neatly arranged wing spots on the underside help to separate Small Blues from Holly Blue – when the wings are open, though, there is no confusion. The two species do also occur in different habitats, in the main.

This is our smallest butterfly – its wingspan barely 2cm. It has dark black-brown uppersides, with a hint of blue near the body. The undersides are light beige with small, white-circled dark spots but no markings along the edges. It can thus be confused with the Holly Blue, but that has a much bluer underside, and its dark spots are not white-circled. The Small Blue has a very patchy distribution – it is found quite widely in southern and central England and south Wales, but otherwise has isolated populations in far south-west and north-west England, coastal east and north-east Scotland, and here and there around the Irish coast.

How to find

- **Timing** In England, Small Blues have a spring and a summer generation, flying from May into June and again in late July to August – the summer brood is typically much smaller. In Scotland, just one generation flies, from June into July.
- **Habitat** It can be found in open grassy habitats of all kinds, wherever the larval foodplant (Kidney Vetch) is abundant.
- **Search tips** Because it is so small and rather dull in colour, this little butterfly is easily overlooked and you should search slowly and carefully, looking close at hand for males perched alert on grass stems or leaves of low bushes, or flitting low over the grass. The Small Blue often basks, turning on the spot to catch the best angle of sunshine, and these movements may draw attention.

WATCHING TIPS

Small Blues are extremely charming in their tiny perfection and make delightful photographic subjects when they are freshly emerged. Males are always on the lookout for females but are not aggressive to their own kind, so you may see them perched in small groups. An ovipositing female, rather dwarfed by the broad Kidney Vetch flowerhead, lays a single egg with care and then rubs her abdomen on the flowerhead – this is thought to leave a scent that discourages other females from laying on that same spot (as the caterpillars are cannibalistic).

Lycaena phlaeas

Small Copper

Wingspan 3cm

1	2	3	4	5

J F M A M J J A S O N D

The Small Copper is not easy to confuse with any other species once you have a good look at it. The upperside forewings are fiery orange with a dark edge and a few dark spots – the hindwings are dark with a band of orange near the margin. The underside hindwing is pale plain brown, the forewing light orange with dark spots and a mid-brown border. This contains a row of shining blue spots in the fairly common aberrant form *caeruleopunctata*. Many other colour anomalies have been recorded, including striking forms in which the orange is replaced with white. The Small Copper is extremely widespread in Britain except the highest parts of Scotland, and is also widespread in Ireland (subspecies *hibernica*).

The variability of this butterfly made it a great favourite with collectors in Victorian times – and with butterfly photographers today. In flight it can look rather drab, but at rest reveals a fiery tint unmatched by other small butterflies.

How to find

■ **Timing** In most years there are two generations of this butterfly, but three is not unusual and occasionally there are four. Therefore, Small Coppers can potentially be seen on the wing any time from mid-April through to mid-October.

■ **Habitat** It occurs in all kinds of open habitats, from downland to heath, well-vegetated shingle foreshores and lake edges, waste ground and larger gardens. The main larval foodplants are Common Sorrel and Sheep's Sorrel.

■ **Search tips** Small Coppers are rarely found in large numbers but most butterfly walks in open country will produce at least one or two. They are rather inconspicuous in their rapid dashing flight but will often stop and (unless it is very warm) immediately open their wings, becoming very eye-catching. They often bask on stones and bare ground and visit ragworts, thistles and other flowers for nectar.

WATCHING TIPS

This is a showy little insect, frequently pausing to bask and to take nectar and turning from one position to another, showing off its bright colours from all angles. Males are territorial and spend time perching on elevated spots with wings half-open, ready to intercept females and to chase off other insects. You have more chance of finding an unusually coloured or patterned individual of this species than probably any other butterfly, so when you are in a good Small Copper area look out for any oddities.

Marbled White

Melanargia galathea

Wingspan 5.5cm

1	2	3	4	5

J F M A M J J A S O N D

One of our most striking butterflies, the Marbled White is a joy to see on a warm summer afternoon in a flowery field. It is part of the group known as 'browns', but its wings have a bold pattern of black and white patches on the upperside. The underside is whiter, with grey markings (light brown in the female). It is mainly found in south and central England but its range stretches patchily through to Cornwall and south Wales, and up to Lincolnshire. It is also on the increase, establishing new colonies and expanding its range northwards and westwards.

Female Marbled Whites are sepia-toned on their underside, while males are grey. This makes for a striking contrast when you see a mating pair – and Marbled Whites are not shy when it comes to their mating activity.

How to find

- **Timing** Marbled Whites are on the wing from mid-June to mid-August.
- **Habitat** They are found in flower-rich grassland areas, including open downland slopes but also more sheltered level ground, such as on woodland edges and even on roadside verges. The larval foodplants are Yorkshire fog and various other common grasses.
- **Search tips** Make an early start to see Marbled Whites basking with wings open in the first sunshine of the morning. Later in the day you should see them flying in a rather leisurely, fluttering manner across their habitat, often stopping to visit flowers. Stake out any stands of knapweeds or thistles and you are likely to have good views of visiting butterflies. Mating pairs often also perch in exposed, obvious positions.

WATCHING TIPS

Marbled Whites are easy and entertaining to watch all through the day. After warming up with a spell of open-winged basking, they roam the habitat in search of flowers and, as the day goes on, the opposite sex. Copulating pairs make attractive photographic subjects, with the contrasting colours of male and female, and their tendency to perch in clear view. Females with eggs to lay do not settle but just drop them in flight while flying over grassy areas. Look out for Marbled Whites away from known sites, and notify your local recorder of any new colonies you discover.

Hipparchia semele

Grayling

Wingspan 5.6cm

| 1 | 2 | 3 | 4 | 5 |

J F M A M J J A S O N D

Although a large butterfly, the Grayling is not easy to see, thanks to its excellent camouflage and rather furtive behaviour. It is patterned in shades of grey, brown and black, and there are two equally large eyespots, one above the other, within a band of yellow-orange on the forewing on both upperside and underside. The underside pattern is very intricately mottled. The Grayling is found around coastlines of most of Britain and Ireland but is much rarer inland. It is represented by six named subspecies, which differ slightly in size and pattern. For example, the subspecies *thyone*, found only on the western side of Great Orme Head in Wales, is much smaller, with a wingspan of 4.8cm.

It may not be the prettiest butterfly, but the intricate camouflaged patterning on the Grayling's wings is a real marvel of natural selection. It is also one of our most genetically varied butterfly species.

How to find

■ **Timing** Graylings fly from late June or early July into September, with some geographic variation in first emergence time.

■ **Habitat** This is a butterfly of open, warm and dry habitats with much bare ground, such as heaths, sand dunes and waste ground. The larval foodplants are a variety of grass species, including Red and Sheep's Fescue.

■ **Search tips** You are most likely to find a Grayling by accident, flashing open its wings then taking off from almost under your feet. Watch it carefully to see where it lands. Spotting the butterfly resting on the ground is difficult, especially as it rests with wings closed and angles its body to the light in such a way that its cast shadow is very small. They are not especially avid nectar-feeders, but males are territorial and will take flight to chase other insects, while females may be spotted in fluttering flight when they are looking for egg-laying sites.

WATCHING TIPS

Graylings reward observation, showing some interesting behaviours if you are lucky. Early in the flight season you could see a male courting a potential mate with a ritualised wing-flicking dance, after which the pair copulate. She may later be seen searching grassy clumps for places to lay her spherical white eggs. Both sexes visit flowers for nectar at times and while feeding may show the wing uppersides. For the real butterfly aficionado, studying the variation in this species at its various different colonies is quite fascinating.

Speckled Wood

Pararge aegeria

Wingspan 5cm

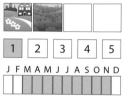

1	2	3	4	5

J F M A M J J A S O N D

One of the most familiar British butterflies, the Speckled Wood is a regular visitor to larger gardens and can be very common in woodland, despite its territorial nature. It has dark brown uppersides with prominent cream-coloured blotches, a single dark eyespot on the forewing (more noticeable on the paler female) and a row of white-circled black spots on the hindwing margin. The underside hindwing is pale brown and cream with a more marbled look and a row of white-centred dark spots. The Speckled Wood is found throughout Ireland, Wales, and most of England, and then discontinuously in north Scotland, with the north-west of Scotland holding a different, paler subspecies (*oblita*).

This characterful, smartly marked butterfly will keep you company on any woodland walk on warm sunny days between mid-spring and mid-autumn. Its long season makes it both the first and the last brown butterfly that you're likely to see each year.

How to find

■ **Timing** Adults fly from late March to October, in a series of overlapping generations (this is because, unlike any other British butterfly, it can overwinter both as a caterpillar and a pupa). Subspecies *oblita*, though, has two distinct generations, flying in May to June and then late July into September.

■ **Habitat** These butterflies are found in woodland, even quite shady spots, clearings, glades, along hedgerows and in parks and larger gardens. The larval foodplants are grasses, including cocksfoot and Yorkshire fog.

■ **Search tips** Any brown butterfly that you notice flying in woodland in somewhat shaded conditions is quite likely to be this species. Look out for it dancing along woodland paths, often pausing to bask in a sunlit spot. It can also be found in more open spaces but is really a woodland specialist, and it will also fly on overcast days.

WATCHING TIPS

Speckled Wood males are alert and territorial, setting out to intercept any passing butterfly. If this is a rival male, or an unwilling female, then an energetic spiralling chase begins, with the male eventually returning to the same or a nearby perch. Females lay their eggs in ones or twos on the underside of grass blades. This butterfly is not a big nectar-feeder but females in particular will sometimes visit brambles and other woodland flowers.

Meadow Brown, Ringlet and Gatekeeper

Meadow Brown

Wingspan Meadow B 5.2cm
Ringlet 4.8cm
Gatekeeper 4.2cm

| 1 | 2 | 3 | 4 | 5 |

J F M A M J J A S O N D

Ringlet

Gatekeeper

These species are very common and widespread. The Meadow Brown is large and brown with some orange on the forewing upperside (very little in males) and a single black, white-centred eyespot within the orange patch. The Ringlet's brown wings bear a row of bold, cream-circled, white-centred black spots on the underside. The smaller Gatekeeper has extensive orange on the upper wings, one large forewing eyespot, and a variegated underside pattern. All are variable, with aberrant forms relatively frequent. The Meadow Brown and Ringlet occur almost throughout Britain and Ireland, while the Gatekeeper is presently confined to England, Wales and southernmost Ireland, but is spreading north.

How to find

■ **Timing** All have a single generation each year. Meadow Browns fly from the end of May into August, Ringlets through June and July into August, and Gatekeepers from mid-July into August.

■ **Habitat** All lay their eggs on grasses of various common species, so are found in grassy habitats: meadows, downland, set-aside fields, roadside verges, large parks and gardens, and grassy edges of other habitats. The Ringlet has a preference for damp ground while the Gatekeeper prefers some bushes or trees.

■ **Search tips** All three are noticeable with their rather lazy fluttering flight and frequent visits to flowers for nectar – they all also bask often, resting on an elevated leaf or (in the case of Meadow Brown particularly) on bare warm ground.

WATCHING TIPS

Because they are so abundant, these butterflies are easy and rewarding to watch. None are particularly territorial, so populations can be very dense and you may well witness males seeking and approaching females and perhaps copulation and egg-laying. Although not colourful they have their own discreet beauty and their willingness to pose at length with wings open makes them great photography subjects. All three species may visit the garden if you have some good nectar-rich flowers and long grass – the Gatekeeper is probably the most frequent garden species.

Small and Large Heaths

Coenonympha pamphilus, C. tullia

Small Heath

Wingspan SH 3.5 / LH 3.8cm

1	2	3	4	5

J F M A M J J A S O N D
Small Heath

J F M A M J J A S O N D
Large Heath

Large Heath

These two related species are the palest of the 'browns', giving a straw-coloured impression in flight. The Small Heath, found throughout Britain and Ireland, is common, if easily overlooked. It always holds its wings closed at rest, showing a pattern recalling that of the Meadow Brown but in a much paler palette of light orange, grey and cream, with a single small eyespot on the forewing. The Large Heath is restricted to northern Wales, northern England, Scotland and Ireland. It shows similar colours but has a row of black, white-centred spots on the hindwing, except in the very pale *scotica* subspecies of north Scotland, in which the main eyespot is reduced to a tiny dot and other spots are absent.

Super sites

Large Heath
- ★ Cors Goch, Anglesey
- ★ RSPB Lake Vyrnwy, Powys
- ★ Whixall Moss, Shropshire
- ★ Threepwood Moss, Scottish Borders
- ★ Glen Nevis, Scottish Highlands

How to find

■ **Timing** Small Heaths fly in two generations in the south, spanning mid-May to early September, and just one further north, from mid-June to mid-August. The Large Heath's single generation flies from mid-June to mid-July – a few weeks later in the far North.

■ **Habitat** The Small Heath occurs in all kinds of open grassy habitats, most commonly on grasslands with a short sward and plenty of sunshine. Its larvae feed on many kinds of grasses. The Large Heath's preferred larval foodplants are cottongrasses, and accordingly it is found on wet boggy moorland where these grasses thrive.

■ **Search tips** Both of the heaths are rather unobtrusive, flying low to the ground and settling with wings closed, half hidden among grasses. They are also easily disturbed. On a slow walk, follow any likely-looking butterfly seen at a careful distance, so you can approach when it settles. Take particular care when searching for the Large Heath, as the ground may be wet or unsafe underfoot.

WATCHING TIPS

These butterflies are tricky to watch, but do visit flowers for nectar. The Large Heath favours cross-leaved heath, while the Small Heath often visits low-growing flowers such as trefoils. The sexes seek each other out when freshly emerged, and join with little obvious courtship. If not disturbed a pair may remain 'in cop' for an hour or more. Females laying eggs can be easier to track as they move more slowly. Large Heath habitat can be good for other rare insects, including White-faced Darter and other upland dragonflies.

Erebia aethiops, E. epiphron

Scotch Argus and Mountain Ringlet

Wingspan Scotch A 4.8cm
Mountain R 3.5cm

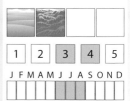

1	2	3	4	5

J F M A M J J A S O N D

Scotch Argus

Mountain Ringlet

These two closely related species are much sought after by lepidopterist visitors to Scotland. While the Scotch Argus is quite easy to find, the Mountain Ringlet represents a great challenge. The Scotch Argus is widespread in Scotland, especially the north and west, while the Mountain Ringlet is confined to the north-western uplands. Both species also have small outposts in the Lake District. Both are very dark brown with a deep orange band on the uppersides of all wings, containing rows of black eyespots, but the Scotch Argus has much stronger markings. The underside of the Mountain Ringlet is plain but the Scotch Argus has a striking silvery band on the hindwing.

Only the most committed lepidopterist will succeed in finding Mountain Ringlet - even in perfect habitat, bad weather can make it impossible to find for days on end.

Super sites

★ Glen Nevis, Scottish Highlands

How to find

■ **Timing** These two species' single generations overlap a little. The Scotch Argus is on the wing from mid-July to early September – Mountain Ringlets fly from June to July.

■ **Habitat** The Scotch Argus occurs in sheltered, damp grasslands in both lowlands and uplands, with blue or purple moor-grass (its preferred larval foodplants). Those in England inhabit dry grassland at woodland edges. The Mountain Ringlet is found on sheltered pockets of damp grassland on mountainsides, with matgrass present (its only larval foodplant).

■ **Search tips** Both species need warm sunny days to fly well, weather conditions that can be hard to find in north-west Scotland. Mid-July offers your best chance of reasonable weather as well as catching both species. Males fly low over the grass searching for females, and both sexes spend much time resting with wings spread to soak up warmth. The Mountain Ringlet's habitat means that you'll need suitable footwear for fell-walking.

WATCHING TIPS

These butterflies are both confiding and easy to watch at close range – the main challenge really is getting to the right spot on the right day. Females of both species are not very active and spend much time hidden in the long grasses, but will bask in the open at times and can usually be inspected at close range. Males on their mate-searching flights will be distracted by anything brown, which you can use to your advantage to encourage them to pause for a while.

Wall Brown

Lasiommata megera

Wingspan 5cm

1	2	3	4	5

J F M A M J J A S O N D

This is one of the most attractive and striking 'browns', with bright orange-brown wings marked with a bold lattice-like dark pattern (heavier in the male). The forewing bears one large eyespot, the hindwing a row of smaller ones. The underside hindwing is mottled grey without bold spots, rather like that of Grayling. The Wall Brown occurs throughout most of England and Wales, thinning out in central northern England, but extending into south Scotland along the coasts. It is also widespread in Ireland. However, this picture masks a serious long-term decline, with many colonies disappearing in all regions, especially away from coastal areas.

When seen in flight, this butterfly could be mistaken for a fritillary with its rich orange coloration. When settled, the big eyespots make it easily identifiable. It is a sun-loving species and basks often, but can be easily disturbed.

How to find

■ **Timing** The Wall Brown has two generations, flying through May and June, and then from mid-July to mid-September. In the North, flight times are a week or so later.

■ **Habitat** It prefers warm, dry and open grassland with bare ground, and is most often found near the sea, on clifftops, the edges of beaches, and waste ground. The foodplants are various common grasses, including cocksfoot, bents and Yorkshire fog.

■ **Search tips** You may spot this flighty butterfly basking on the ground or, often, a wall, though it will usually be quick to take flight when you approach. Look for it on warm sunny days, moving slowly through the habitat and particularly checking any bare surfaces facing the sun. It will also visit flowers – check clumps of knapweeds, scabious and thistles.

WATCHING TIPS

This butterfly is active and nervous but will return to preferred basking spots if undisturbed. Males stand guard and fly out and chase other passing butterflies – if a receptive female comes along the pair quickly tumble down to the ground where they mate. They will also visit nectar-rich flowers of all kinds, and here females may tolerate a closer inspection. Also watch for females looking for egg-laying sites, often on grass growing in small sheltered hollows.

Laothoe populi

Poplar Hawkmoth

Wingspan 7.5cm

1	2	3	4	5

J F M A M J J A S O N D

A very large hawkmoth, this species is not very colourful but has a unique and distinctive shape when at rest, with the top edge of the hindwings showing above the swept-back forewings. The lower part of the hindwings, normally covered by the forewings, is bright orange-red, so the moth can reveal a startling patch of colour if it is disturbed. The wings are otherwise marbled light grey and light brown, with some dark bars and a vague white spot near the centre of the forewings. It has a large bulky thorax but a relatively slim abdomen, the tip of which is often curled up. This moth is widespread in Britain and Ireland, and most common in southern England.

The sudden arrival of a Poplar Hawkmoth in your front room on a warm night can be alarming, but this is a placid creature which will not usually object to being caught and handled. Be careful picking up this gentle giant and avoid touching its wings.

How to find

■ **Timing** Adults fly at night from May to July.

■ **Habitat** This moth occurs in woodland, parkland and large gardens, where its larval foodplants (poplar, aspen and sallow) grow.

■ **Search tips** This moth is usually found when it is attracted to an outside light and is discovered resting on a wall nearby the following morning. Finding it in a more natural setting is difficult as it is well camouflaged against the tree trunks where it rests. It will come to moth traps.

WATCHING TIPS

Like other hawkmoths this is a spectacular creature that is usually happy to be gently handled, so if you find one you can look at it at length before letting it go in a well-hidden place. The horn-tailed caterpillars are variable in colour – those that feed on poplar leaves are more blue-green in tone while those feeding on sallow are more yellowish. They may be encountered in autumn on the ground, searching for somewhere to pupate.

Eyed Hawkmoth

Smerinthus ocellata

Wingspan 7.5cm

1	2	3	4	5

J	F	M	A	M	J	J	A	S	O	N	D

This moth is unmistakeable if it shows its pink hindwings with their beautiful large eyespots, which are blue with a dark centre and border. The spots are flashed when the moth is alarmed, but otherwise covered under the pink-grey and brown forewings. Like this, it could be confused with the more colourful Lime Hawkmoth, but a quick way to identify it is by the broad black line down the centre of the thorax – this stands out as the darkest part of the moth. It is widespread but rather patchily distributed in England and Wales becoming scarce in north England and central Wales. It is also present in Ireland but does not appear to be common there.

The larva has the typical hawkmoth spike at its tail end.

How to find

- **Timing** Adults are on the wing at night from May to July.
- **Habitat** This is a moth of wooded areas, and it is a fairly frequent visitor to larger gardens. The larvae feed on foliage of willow, sallow and other trees, and may be found on garden apple trees.
- **Search tips** Look for the Eyed Hawkmoth resting on walls, wooden fences and tree trunks. It is attracted to light and so will come to moth traps and may be found early in the morning after a warm night, resting close to an outdoor light source.

WATCHING TIPS

This moth is easy to handle and may flash its eyespots at you, revealing its full beauty. If you are lucky enough to find an Eyed Hawkmoth larva on your garden apple tree, you can try rearing it – the mature caterpillar is green with white stripes and a short blue tail horn. As with other hawkmoth caterpillars it will need some soft loose earth in which to pupate and will need to stay outdoors through winter. The pupa is blackish and glossy, and will be formed just under the soil surface.

Macroglossum stellatarum

Hummingbird Hawkmoth

Wingspan 4.5cm

1	2	3	4	5

J F M A M J J A S O N D

Even the most committed moth-phobe could not help but be impressed by this remarkable little hawkmoth as it hovers expertly around a clump of flowers in an upright posture, looking uncannily like a tiny hummingbird. In flight the brown and rufous wings are a blur but the long black proboscis is evident, as is the black-and-white banded abdomen tip. This species is a migrant to Britain and in very good years significant numbers will reach northern England and even Scotland, but it is always most numerous in southern and eastern regions, especially on the coast.

This remarkable little moth has an extremely fast metabolic rate to fuel its 85 wingbeats per second – studies suggest that its energy exchange system operates on the very limit of what is biologically possible.

How to find

■ **Timing** The first Hummingbird Hawkmoths are usually seen in May, and more appear through summer, when some locally-bred new adults join them. Arrivals are most likely in spells of warm weather with winds from the South. They are diurnal.

■ **Habitat** This moth is a great traveller and could turn up anywhere, but is constantly in search of flowers, especially those with clustered small blooms, such as red valerian and buddleia – it is often seen in gardens. It also seeks out bedstraw, the larval foodplant.

■ **Search tips** Keep a lookout on wildlife-watching social media for news of sightings as spring progresses. Butterfly Conservation's Migrant Watch page keeps track of records of this and other migrant lepidoptera. If conditions look promising, search for it in sunny places with lots of flowers, ideally near the coast.

WATCHING TIPS

This is a delightful moth to watch. Although it is not our only hawkmoth that feeds in flight, it is the most likely one to be seen doing so by day. The precision of its movements as it hovers and feeds is quite astonishing. Sadly, it is also prone to disappear in an instant, but if you are lucky it will linger. When perched it looks very different, with the orange hindwings hidden – check for the flared-out black-and-white abdomen tip. Plant buddleia and valerian to attract it – these will of course appeal to many other nectar-feeding insects too.

Moth-trapping

Moths are famously drawn to light sources, and this provides wildlife-watchers with opportunities to catch and study them. Modern moth traps are sophisticated and reasonably affordable, allowing anyone to gain a much deeper insight into the diversity of moth species found in their gardens, and to see species that are very difficult to find in natural conditions. Of course, traps must be used ethically, monitored carefully, and all caught moths must be released in safe places the following day.

Designs

Lights for moth traps are designed to transmit a wider range of wavelengths than a standard household bulb – in particular, ultraviolet light. Mercury vapour bulbs and actinic strip lights both produce the correct wavelengths. In a standard moth trap, the light is placed on top of a large collecting box, with a clear lid that has an opening in the middle. Inside the box, there are usually pieces of broken eggbox, giving the moths somewhere to sit and rest. The light is powered from the mains or (if you are using it a long way from a power supply) a car battery or a portable generator. You can make your own, but unless you are expert with electricals it is wise to use a commercial model, as inadequate weather-sealing will make the device dangerous.

Using light traps will permit good views of beautiful species like this **Pale Tussock**.

Moth traps using pheromone lures, to attract males of particular species, can also be purchased or home-made. It is even possible to make a trap housing an unmated female moth – this is a popular way of attracting Emperor Moths, for example. To find out about making pheromone traps, visit angleps.com/pheromones and follow the links.

This **modern trap design** includes its own battery, so is economical and highly portable.

Why light?

Over the centuries, people have noticed how moths will fly to light, with fatal results if the light in question is a candle. This apparently foolish behaviour stems from a sensible natural behaviour – in centuries past, the only light source at night would be the moon, a fixed point which moths could use as a navigational aid by maintaining a constant position relative to it as they flew. Of course, to achieve the same with a much closer light source will result in the hectic circling behaviour we see when a moth flies to a light.

Not all moths are drawn to light, but many are. The standard moth trap exploits this behaviour, drawing moths close with an electric light source. Their circling deflects them into a catching container from which escape is difficult, so most that are caught will stay caught.

An impressive and varied **moth trap haul**. High summer is the time of peak moth diversity.

Dealing with the catch

You may choose to sit up with your trap and watch the moths as they arrive. If not, inspect it first thing in the morning. Most moth-trappers will keep a record of species and numbers of individuals caught – over the weeks, months and years this will provide useful data as to what is about and when. If you are interested in photography, you can place your catches in good settings for photos too, but do move them to another spot out of sight when you have finished, rather than leaving them in full view where predators could easily take them.

Moth-trapping in the garden can be very rewarding but **if you can go further afield** (with the land-owner's permission) you may find scarcer species.

When handling moths, be extremely gentle. By day, most moths are sluggish and can easily be persuaded to climb onto a finger if you gently touch their legs. Try to avoid touching the wings and body as this can rub off the scales.

Occasionally, something untoward happens in the moth trap. You may catch wasps, even Hornets, which should obviously be released with great care – they may also kill and eat moths in the trap. A happier event is when a trapped moth lays eggs in the trap. If you feel so inclined, you can attempt to keep the resultant caterpillars until they pupate and the adult moths emerge, provided you can provide the caterpillars with the correct larval foodplant. Wildlife-watching suppliers such as NHBS (nhbs.com) sell containers suitable for rearing caterpillars, as well as moth traps.

Elephant Hawkmoth

Deilephila elpenor

Wingspan 5.2cm

1	2	3	4	5

J F M A M J J A S O N D

The caterpillar has distinctive eye-like markings behind its head.

This is a spectacular hawkmoth, giving the lie to the idea that moths are drab creatures with its intense pink coloration, most vivid on the underside of the body. The forewings are sleek, with smooth rather than scalloped edges, and patterned in pink and moss green – the hindwings are pink with black bases. The abdomen is mainly green on the upperside, and plump but strongly tapered with a pointed tip. The striking caterpillar, brown or green with a bulbous head-end bearing large false eye markings, is seen as often as the moth, if not more so. This moth is widespread in Britain and Ireland, with its densest populations in the South and East Anglia.

How to find

▪ **Timing** Elephant Hawkmoths are on the wing at night from May to July. Through August into September, the mature caterpillars are often encountered in the daytime as they leave their foodplants and search for pupation sites, sometimes even taking time to bask in sunshine.

▪ **Habitat** This species is found in woodland, gardens, and other habitats with lush herbaceous vegetation. The main larval foodplant is Rosebay Willowherb but it will also use garden fuchsias.

▪ **Search tips** The adult moths are strongly attracted to light so may be found perched on walls of buildings in the early morning. At more natural roosts such as branches with patches of green lichen, they can be surprisingly well camouflaged.

WATCHING TIPS

This is a real prize to catch in a garden light trap, and seems to be on the increase. The wandering caterpillars could be persuaded to pupate in a container with some loose earth, giving the opportunity to see the freshly emerged moth the following spring. This species could be confused with the Small Elephant Hawkmoth, though that species is much smaller, with blunter wings and paler, less intense coloration.

Sphinx pinastri

Pine Hawkmoth

Wingspan 7cm

1	2	3	4	5

J F M A M J J A S O N D

A dark, sleek, somewhat imposing-looking hawkmoth, this species is closely tied to pine forest. The long, broad forewings are grey with some dark streaks, affording good camouflage against Scots pine bark. There are two distinctive black stripes either side of the thorax. The caterpillar is bright green with yellow longitudinal stripes, and with maturity it also develops a row of red spots along its sides. This moth is most common in southern England and East Anglia. Its range extends very patchily south-west to Cornwall, and towards north-east England.

The caterpillar is more colourful than most hawkmoth larvae.

How to find

■ **Timing** Adults are on the wing at night from May to July, or just into August.

■ **Habitat** This moth lays its eggs on pine needles, primarily Scots pine. It is therefore most often found in coniferous woodland, but could occur in mixed woodland as long as some pines are present.

■ **Search tips** Like most other hawkmoths the Pine Hawkmoth is attracted to light, and could turn up at a trap if you are close to suitable habitat. When walking in pine woodland in late spring or early summer, take the time to check tree trunks for the well-camouflaged adult.

WATCHING TIPS

This moth feeds on nectar at night, making an impressive spectacle as it hovers in front of night-scented flowers such as honeysuckle and jasmine. You might see this behaviour if you go in search of it on a well-lit, still and warm night, along the edges of pine woods. Although this is a rather drab hawkmoth colourwise, its large size and sleek fighter-plane shape make it an impressive beast, and a great 'find', whether in the woods or at a trap.

Broad-bordered Bee and Narrow-bordered Bee Hawkmoths

Hemaris fuciformis, H. tityus

Wingspan Broad B 4.5cm
Narrow B 4cm

1	2	3	4	5

J F M A M J J A S O N D

Narrow-bordered Bee Hawkmoth

Broad-bordered Bee Hawkmoth

These small hawkmoths, which fly by day and feed in flight, are mimics of bumblebees. They have clear wings, the front pair much larger than the hind, and short, rounded, fluffy bodies. The Broad-bordered has a wide red border to all wings, and red wing veins, dark tail sides and a red band across the otherwise light fawn abdomen. The Narrow-bordered has a thin black edge and dark veins to all wings, and its abdomen is patterned in black, rufous and cream. Both species are rather uncommon. The Narrow-bordered has a wide but very patchy distribution across Britain and Ireland. The Broad-bordered is confined to south and east England, but, within its range, is more common than the Narrow-bordered.

These two species are examples of Batesian mimics – harmless species that resemble another, potentially dangerous species. Their shape and colour mimics that of a bumblebee and discourages predators that have already encountered real bumblebees.

How to find

▪ **Timing** Both of these species are on the wing in May and June, and are diurnal.
▪ **Habitat** The bee hawkmoths occur in woodland clearings and edges, and other sheltered, often damp and lush habitats, offering plenty of wild flowers. The Broad-bordered uses scabiouses for its larval foodplant, while the Narrow-bordered's main foodplants are honeysuckle and bedstraw.
▪ **Search tips** These moths are rather small and tend to feed from low-growing flowers, so are easily overlooked. They are also unpredictable in their appearances. You are most likely to spot one if you sit down in a suitable patch of habitat, with lots of flowers such as trefoils and bugle, and keep an eye on the flowers. After the flight season, look out for the Broad-bordered's larval feeding signs – neat circular holes in honeysuckle leaves.

WATCHING TIPS

These moths are agile feeders, impressive to watch as they manoeuvre around flowers on fast-buzzing wings. They position themselves closer to the target flower than the Hummingbird Hawkmoth does but have similar aerial skill. Identifying them is tricky if they don't settle, but because they hover on the spot it is possible to photograph them in flight if you have a good camera – identification from the photos should then be possible. Moth recorders will be interested in sightings from all regions as neither species is common.

Mimas tiliae

Lime Hawkmoth

Wingspan 6.2cm

1	2	3	4	5

J F M A M J J A S O N D

An attractive hawkmoth, this species is a common visitor to gardens, especially in areas where avenues of lime trees are planted. It has narrow wings with scalloped edges, and in general shape recalls the Eyed Hawkmoth but is more colourful, being soft pink with a darker green band across the centre of the forewing, and dusky forewing edges with white patches at the tips. A scarce brown form exists. The thorax has a pale centre and broad dark stripes on either side. Mature caterpillars are a distinctive pink-grey colour. This species is found mainly in southern and central England but is spreading north.

The caterpillar bears a prominent thorn-shaped spine on its rear end.

How to find

■ **Timing** Lime Hawkmoths fly at night in May and June. In late summer the mature caterpillars leave their foodplants and seek a place to pupate.

■ **Habitat** This moth occurs in deciduous woodland but also in parks and gardens where there are plenty of lime trees (the main larval foodplant – it also uses birches, elms and alder).

■ **Search tips** This moth is attracted to light. You may also find it resting by day on a lichen-dappled wall or tree trunk. Also look out for the full-grown caterpillars later in summer, as they walk over the ground on their way to pupate.

WATCHING TIPS

These attractive moths often turn up in moth traps, and around outside lights – they are easily handled when found in daylight so you can examine their gorgeous colours. Sightings in northern and western counties should be passed on to the local moth recorder. It is not unusual to unearth Lime Hawkmoth pupae in the garden in winter and spring - if you do, rebury them just at the soil's surface. Alternatively, relocate to a pot (again, bury in shallow loose soil), check the pot each night from early May, and you may see the adult moth when it emerges.

Privet Hawkmoth

Sphinx ligustri

Wingspan 10.5cm

1	2	3	4	5

J F M A M J J A S O N D

The colossal caterpillar is commonly found on or near privet hedges.

A huge moth, this is the biggest species regularly encountered in Britain and is not infrequently found in gardens. Its caterpillar is, if anything, even more impressive when fully grown. The adult is dark with long, narrow, pointed forewings. They are dark grey on the rear edge, paler creamy-grey on the leading edge, while the hindwings are striped pink and black. The stout abdomen is striped bright pink and black with a grey central stripe, while the thorax has pale sides and a black upperside with a slight sheen. Privet Hawkmoths are widespread across south and south-west England and East Anglia, and occur patchily in Wales, the Midlands and northern England.

How to find

■ **Timing** This moth is on the wing in June and July, at night. The spectacular mature caterpillars can be found in late summer.

■ **Habitat** It is found in woodlands, parks and gardens, wherever the larval foodplants (primarily privet, but also lilac and ash) grow.

■ **Search tips** Adults come to light. If you have a privet hedge it is worth looking for the large caterpillars later in summer, as they feed quite openly on the foliage.

WATCHING TIPS

The chance to examine this magnificent moth at close range is reason enough in itself to invest in a moth trap, or to take the time to raise a caterpillar if you find one. If you know the species is present in your garden, look out in autumn or winter for the very large, glossy black-brown pupae, buried in shallow earth or leaf-litter. Transfer pupae to a container (but leave outdoors) for a chance of seeing the moth emerge the following summer. The similarly huge Convolvulus Hawkmoth, which resembles this species, is a rare migrant to Britain but may turn up in late summer.

Arctia caja

Garden Tiger

Wingspan 5.5cm

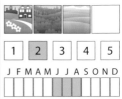

| 1 | 2 | 3 | 4 | 5 |

J F M A M J J A S O N D

This is one of the furriest caterpillars you'll see in the British Isles.

A very distinctive moth in both its adult and larval forms, the Garden Tiger was very common a few decades ago but has declined substantially, although its distribution through Britain and Ireland remains extensive, particularly around the coast. The broad, round-tipped forewings bear large dark brown blotches on a white background, forming an attractive swirling pattern. The hindwings are bright red with large, black-bordered dark blue spots. The thorax is dark brown, the abdomen red with dark central markings. The caterpillar has a luxuriant coat of long hair which is red on its sides and black on the top, with white tips that create a silvery look.

How to find

■ **Timing** Adults fly at night in July and August. The fluffy 'woolly-bear' caterpillars may be seen walking purposefully (and quickly) over the ground in early summer, in the daytime, as they look for places to pupate.
■ **Habitat** This moth can be found in larger, well-vegetated gardens, parkland and woodland glades and edges. It uses many different larval foodplants, including Common Nettle, bramble, honeysuckle and heathers, as well as some cultivated garden plants.
■ **Search tips** The adults are nocturnal so hard to find, but are attracted to light and may be found perched on fences and walls close to lit buildings. The caterpillars are most often discovered crossing lawns or pathways.

WATCHING TIPS

Many lepidopterists mourn the loss of this species as a ubiquitous garden visitor. Allowing wild patches in your garden where herbaceous plants can proliferate will encourage it. Avoid handling the caterpillars, charming though they are, as their hairs can cause a skin reaction in some people. A few related species also have luxuriantly furry caterpillars, but the combination of silver-tipped black hairs and the reddish side-stripes make this species easily identifiable.

Jersey Tiger

Euplagia quadripunctaria

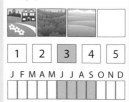

Wingspan 5.2cm

1	2	3	4	5

J F M A M J J A S O N D

Another striking tiger-moth, this species is easily recognised by the cream vertical tiger stripes on its black forewings and thorax. The hindwings are red or yellow-orange with large black spots – the abdomen is also red-orange. The Jersey Tiger's name is increasingly inaccurate as it is no longer confined to the Channel Islands and south-west coast but is spreading north and east quite rapidly – it is now fairly common in London and occurs in most other southern counties. Amateur lepidopterists' observations are helping to chart its spread.

This beautiful moth is an increasingly frequent sight in southern England, and each year is found further and further from its original south-western range. In 2017 it was recorded in Hertfordshire for the first time, a new northern limit for the species.

How to find

▪ **Timing** These moths fly from July to September and are active in the day as well as at night.

▪ **Habitat** It is found in a variety of habitats, including quite open and exposed marshland as well as in gardens and on woodland edges. Caterpillars feed on Common Nettles and other kinds of herbaceous wild plants.

▪ **Search tips** You will often find Jersey Tigers feeding on buddleia in high summer, alongside the usual range of butterflies. They will also come to outside lights, and will rest conspicuously in full view, relying on their bold warning coloration to discourage predators.

WATCHING TIPS

This is a stunning moth, particularly when it takes flight and reveals its colourful hindwings, and is easy to watch. If you have a garden buddleia, check for this species on the blooms through summer, even if you are outside of its present range, and let your local moth recorder know if you find it. With its elegant creamy forewing stripes, it warrants the name 'tiger' more than its spot-patterned relatives. It is one of our most striking moths and a very welcome addition to our fauna.

Other tiger moths

Cream-spot Tiger moth

1	2	3	4	5

J F M A M J J A S O N D

Ruby Tiger moth

Tiger moths form part of the family Erebidae. As a group, the tiger moths are among our most attractive and eye-catching moths. Besides the Garden Tiger and Jersey Tiger, other species that are quite common and widespread in Britain are Ruby Tiger, Scarlet Tiger, Wood Tiger and Cream-spot Tiger. All have brightly coloured hindwings (yellow or red), and black forewings with cream or white markings, except for the Ruby Tiger which has red-brown forewings. All have very hairy caterpillars. Most tiger moths reveal bright colours when they take flight – even the rather plain-looking, brownish Ruby Tiger sports a pair of deep crimson hind wings and a fluffy red body.

Scarlet Tiger moths mating

How to find

■ **Timing** These species all fly in spring or early summer, with the Ruby Tiger having a second generation in late summer, and you could see them all on the wing in daylight.

■ **Habitat** Tiger moths are mainly found in lush, sheltered and well-vegetated habitats, such as woodland edges and clearings, sheltered meadows, riversides and edges of marshland. All of them have a variety of larval foodplants, including many common herbaceous species.

■ **Search tips** Some tiger moths do not feed at all in their adult form, but all are attracted to light. They are eye-catching in flight and often rest in the open.

WATCHING TIPS

The bright colours of tiger moths warn predators of their toxic bodies, achieved through eating plants containing noxious chemicals in their caterpillar stage, and their caterpillars are similarly off-putting with their long hair. Most insect-eaters will therefore avoid them, although Cuckoos are specialist feeders on these and other hairy caterpillars. The Scarlet Tiger in particular is often found by day, seeking nectar, and sometimes occurs in impressively high concentrations.

Emperor Moth

Saturnia pavonia

Wingspan 5cm

1	2	3	4	5

J F M A M J J A S O N D

This is the only British member of the family Saturnidae, a group which includes the domesticated silkmoth. The Emperor Moth is a large, unique and very striking moth with a large eyespot highlighting each velvety-looking, beautifully patterned wing. Males have brown forewings and orange hindwings, and sport extravagant feathered antennae, while the much larger females are grey. The mature caterpillar is distinctive too – bright green with rings of raised yellow or pink 'warts', surrounded by black hairs, on each segment. It is a very widespread species but not usually particularly common.

With its stout, cylindrical shape and evenly spaced, warty bands, the mature caterpillar is just as striking as the adult moth, though arguably not as beautiful. Younger caterpillars are black with dense white spines – as they grow the green sections appear and expand.

How to find

■ **Timing** This moth flies in April and May, with males active by day, seeking out the females which sit well-hidden in heather. Later in the season, females fly by night to lay eggs.

■ **Habitat** It is found most commonly in open habitats such as heaths and moors. The main larval foodplants are heathers.

■ **Search tips** Look out for the males flying fast and low when you are out walking in heathery countryside in spring. They fly long distances to track down females, locating them by scent from a mile away or more, even before the female has fully emerged from her pupa. Also look out for the distinctive caterpillar feeding on heather.

WATCHING TIPS

The surest way to see Emperor Moths and witness their most famous behaviour is to find an unmated female or, even better, a pupa containing a female. From the very start she releases pheromones which will attract males from all directions, and will not stay unmated for long – most will have paired within minutes of emergence. In their adult form, neither sex is able to feed, so the whole of their short lives is devoted to reproduction.

Tyria jacobaeae

Cinnabar

Wingspan 3.8cm

| 1 | 2 | 3 | 4 | 5 |

J F M A M J J A S O N D

This colourful moth is often presumed to be a butterfly, and is often seen by day. The forewings are black with two red spots and a red longitudinal line on the leading edge, while the hindwings are red. The body is all-black. It could be confused with the burnet moths, but they are much stouter, with clubbed antennae. The Cinnabar's gregarious caterpillars are very conspicuous with their black-and-yellow bands, relying on their warning coloration to deter predators. The Cinnabar has declined over the last few decades but remains quite common and widespread in England, Wales and Ireland. It is scarcer in north England and very sparsely distributed in Scotland, with most colonies at or near the coast.

The Cinnabar is one of several moths which are probably more familiar in their larval form than as adults. The caterpillars sequester toxins from their food plant in their bodies, making them unappetising for predators, and allowing them to feed in full view.

How to find

■ **Timing** Adults fly from May to July. They are nocturnal but regularly take flight when they're disturbed from long grass in daytime. The caterpillars can be seen all through summer.

■ **Habitat** This moth likes lush green open spaces, such as meadows, woodland glades, downland, marshland edges and similar habitats. The larval foodplant is ragwort.

■ **Search tips** You are most likely to encounter a Cinnabar adult while walking through a field of long grass – if you step too close to its daytime resting place it will take flight, showing its brilliant red hindwings. Adults also come to light at night. The caterpillars are easily found – look in July or August, in any area where ragwort grows in profusion, checking the top parts of each plant.

WATCHING TIPS

When disturbed from its roost, a Cinnabar usually makes just a short flight before landing within the grass again, and if you approach slowly you should be able to get a good look at it. If you live rurally you may also attract it to a garden moth trap. Ragwort is not always tolerated by landowners as it is poisonous to livestock (although they do avoid eating the living plants) – but if you have a large garden and don't mind some ragwort plants you may attract breeding Cinnabars.

Burnet moths

Zygaena

Wingspan *circa* 3.5cm

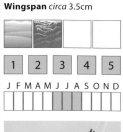

| 1 | 2 | 3 | 4 | 5 |

J F M A M J J A S O N D

Green Forester moth

The burnets are a distinctive group of day-flying moths, often mistaken for butterflies because of their bright colours. Typically, the forewings are glossy black-green with red spots (large patches in the case of the rare Transparent Burnet), and the hindwings are red. Burnets have plump black bodies, and their long, stout antennae are club-tipped. The most common species in Britain are the Six-spot Burnet, found throughout Britain and Ireland, and the Five-spot and Narrow-bordered Five-spot, both quite widespread in southern and central England and Wales. The latter also has a small presence in Scotland and Ireland. The forester moths are close relatives of the burnets and look similar, but have unmarked shiny green wings.

Five-spot Burnet moth

The Forester is a distinctive, burnet-like day-flier with glossy wings and stout antennae.

Super sites

★ Isle of Mull, south-facing grassy slopes (Transparent Burnet, Slender Scotch Burnet)

★ Cairngorms slopes, Scottish Highlands (Scotch Burnet)

How to find

■ **Timing** Burnets fly in high summer, with July being the best month to see most species in profusion. They are diurnal.

■ **Habitat** These are moths of open flowery grassland. The larval foodplants are mainly low-growing members of the pea family – Birdsfoot Trefoil in the case of the Six-spot and Five-spot.

■ **Search tips** Burnets are avid nectar-feeders and are most likely to be spotted on the flowers of knapweeds, thistles, Wild Thyme and similar, often more than one sharing the same bloom. They are active in sunshine – on a dull day, have a look for the empty pupal cases (papery yellow husks attached to plant stems, often with the black remains of the caterpillar's last moult attached) to confirm their presence.

WATCHING TIPS

At the best sites, burnets fly in such profusion that they make for quite a spectacle. It will seem that every flower has a moth in situ, or a pair of moths 'in cop'. To find and watch the rare burnet species will probably require a special trip (see 'Super sites'), and taking photographs is wise to ensure you make a correct identification. All burnet moths are easy and enjoyable to watch as they bumble from flower to flower and engage in spontaneous romantic encounters with others when they meet on flower heads. They can usually be handled quite easily if you encourage them gently onto a fingertip.

Clearwings

Large Red-belted Clearwing

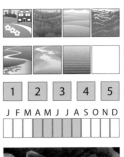

1	2	3	4	5

J F M A M J J A S O N D

Hornet Clearwing

This is a group of distinctly un-mothlike moths, with transparent wings and more than a passing resemblance to flies or other insects. The group also includes two large and very convincing large hornet mimics – the Hornet Moth and Lunar Hornet Moth. The smaller members of the group often have strikingly long abdomens, marked with coloured bands or patches, and with expanded, fluffy abdomen tips. Some clearwing moths are widespread, others localised – the majority are found in parts of southern England, but all are generally rather difficult to see. The larvae typically live inside parts of the foodplant so are almost impossible to see.

Some clearwings are astoundingly convincing hornet mimics – with their very stout striped bodies they are almost more alarming than the real thing. Most species are much smaller, resembling solitary wasps of various kinds.

How to find

■ **Timing** Clearwings mainly fly in the summer months. Some species are diurnal, others nocturnal.

■ **Habitat** These moths are usually found close to their larval foodplant. The Hornet Moth requires poplar or Black Poplar, while the Lunar Hornet Moth's larval foodplants are sallow and willows. Most other clearwings use deciduous trees but some use various herbaceous plants.

■ **Search tips** Look for Hornet Moths and Lunar Hornet Moths resting low on the trunks of their respective foodplant trees. To see other clearwings, the best way is to use a pheromone lure – these can be purchased from entomological suppliers such as Anglian Lepidopterist Supplies: angleps.com/pheromones.

WATCHING TIPS

Many species of moths can be attracted by synthetic pheromones, which mimic the natural chemicals released by female moths to attract males. Therefore, pheromone traps will only bring in males of your target species, not females. Normally the lure is placed inside a cloth bag and hung up in a tree or bush (your lure will come with full instructions) – check often to see what has arrived. Pheromone lures work very well on several species of clearwing moths and are used by moth groups to confirm the presence of these elusive insects in particular areas.

Buff, White and Water Ermines *Spilosoma lutea, S. lubricipeda, S. urticae*

Buff Ermine moth

Wingspan 3.5–4cm

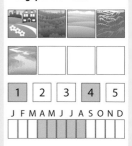

| 1 | 2 | 3 | 4 | 5 |

J F M A M J J A S O N D

White Ermine moth

The ermines are distinctive and attractive relatives of the tiger moths. The White and Water Ermines are mostly white, fluffy moths with black spots on their wings (several in the case of the White, almost none with the Water Ermine). The Buff Ermine is a beautiful rich cream colour, with a row of black spots across its wings. All have black antennae and legs. The hairy larvae live in conspicuous feeding webs. The White Ermine is widespread throughout Britain and Ireland – a dusky form occurs in Scotland. The Buff Ermine is also widespread but very scarce in Scotland. The Water Ermine occurs along south-eastern coasts, and has a few colonies elsewhere in southern England.

Related to the ermines are the Yellow-tail and Brown-tail Moths, which are pure white and fluffy as adults, with coloured abdomen tips, and which feed in communal webs as larvae. Brown-tail larvae are notorious for causing skin irritation if handled.

How to find

■ **Timing** All of these nocturnal moths are active from late spring to mid-summer.
■ **Habitat** Buff and White Ermines occur in a range of habitats, including gardens, and use a wide variety of larval foodplants. The Water Ermine is found on marshes and fenlands, and lays its eggs on water mint and other wetland herbaceous plants.
■ **Search tips** These moths are all nocturnal and are all drawn to light. Look out for them resting on walls or windows that are lit up overnight. The caterpillars are also conspicuous.

WATCHING TIPS

Buff and particularly White Ermines have long been popular subjects for entomologists, as their patterns show great variability. This has been exaggerated through captive breeding to produce some unusual and beautiful patterns. When found by day, ermines are easily handled but should be placed under concealing cover once you have had a look at them. The young caterpillars have interesting communal feeding behaviour, constructing large, sheet-like silk webs for shelter. They should not be handled though as their hairs can cause severe skin irritation.

Lasiocampa quercus

Oak Eggar and other eggar species

Wingspan 6cm

| 1 | 2 | 3 | 4 | 5 |

J F M A M J J A S O N D

Oak Eggar

Fox moth

The Oak Eggar is a large and distinctive moth, found throughout Britain, and one of rather few moths of the family Lasiocampidae that occur in Britain. Oak Eggars are brown and fluffy with a pale band across each wing, and a white spot on the centre of the forewing. Others in the group are similar, being brown or grey, stout, fairly large, very furry and broad-winged. They include the Fox Moth, which has a double white stripe rather than a white spot on the forewing, the Lappet and the Drinker, which are both remarkable leaf-mimics, and the December Moth, a greyer species that flies in winter. Lasiocampid caterpillars are distinctive, with bold patterns formed by coloured bristles.

The lasiocampid moths are stocky and heavy-bodied, with luxuriantly furry bodies and velvety-looking wings. The males tend to have large comb-shaped antennae, which help them sense any newly emerged females nearby.

How to find

■ **Timing** Oak Eggars fly in summer, as do most other species (The December moth is the obvious exception). Male Oak Eggars fly by day, seeking females, while females are nocturnal. With most others in the group, both sexes are nocturnal.

■ **Habitat** The Oak Eggar is most common on heaths and moorland, where its larvae feed on heather and bilberry. The other lasiocampids mostly occur in woodland and scrub.

■ **Search tips** These moths are all attracted to light. Otherwise, they are most easily found in their caterpillar stage. Drinker moth caterpillars are encountered particularly frequently, drawing attention with their conspicuous blue-ish colouration and large size.

WATCHING TIPS

These moths do not feed as adults, so their lives are brief and powered by stored fat from their time as a caterpillar. Their only goal is reproduction, and typically males seek out females, attracted by pheromones that the females release from glands in their abdomen. This means you are more likely to see males than females. Mature caterpillars can be encouraged to pupate in outdoor containers, giving you a chance to see the adult moth in due course (and if it happens to be a female, it is very likely to attract some males).

Magpie Moth

Abraxas grossulariata

Wingspan 3.7cm

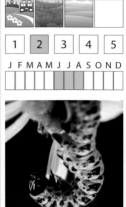

1	2	3	4	5

J F M A M J J A S O N D

This moth is unique and very striking. It is white with the forewings marked with bands of black and orange-yellow spots. It has large wings and a slender body, so is often mistaken for a butterfly. The larva is also distinctive, with colours echoing its adult form. Part of the family Geometridae ('earth-measurers') it has a typical looping motion when it moves. Magpie Moths are declining but are still fairly common and widespread in the South, including in gardens. The Small Magpie is superficially similar to this species but has no orange on the wings, and belongs to a different family. However, the Clouded Magpie is a close relative, and has a similar pattern but with pale grey rather than black spots.

The caterpillar and moth both sport warning colouration, and any predator that ignores this will find the insect (whether larva, pupa or adult) an unpleasantly bitter-tasting mouthful, thanks to the chemical sarmentosin, which it stores in its body.

How to find

■ **Timing** Magpie Moths are on the wing in July and August, and are nocturnal, but are conspicuous enough that they are often seen resting in the daytime.

■ **Habitat** This is a moth of scrubby areas, where its foodplants (currants, gooseberries and related species) grow. It can also be frequent in gardens where these plants are grown.

■ **Search tips** The striking adult moth comes to light, and may also be seen resting on walls or foliage in the daytime. The caterpillars are easily found on the foodplant, as are the pupae, which are striped black and yellow, and attached to the undersides of leaves with silk.

WATCHING TIPS

This moth can be attracted to your garden if you grow gooseberries and currants, and don't mind tolerating the caterpillars' depredations. If you encourage a thriving population, you may in due course see some of the pattern variations that made this moth such a favourite of early lepidopterists. The rarer Clouded Magpie may also occur in gardens but relies on elms as larval foodplants. Its caterpillars have narrow black side-stripes, a pattern most unusual among British lepidoptera.

Swallow-tailed Moth

Wingspan 4.5cm

1	2	3	4	5

J F M A M J J A S O N D

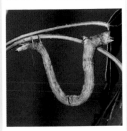

This is a beautiful and unique moth and one of the largest of the family Geometridae to occur in Britain. It has large, very delicate-looking wings with tail-like projections on the hindwings, giving it a shape recalling the Swallowtail butterfly. The wings are pale creamy-yellow, becoming whiter over time, with two fine brown lines running down the forewings and one on the hindwing, and fainter dark markings in between. The caterpillar is brown and twig-like, and often adopts a straight, rigid pose to enhance its camouflage. This moth is found quite widely across Britain apart from north Scotland.

This unmistakable moth is a regular visitor to lit walls and windows through the height of summer. Its well-camouflaged, twig-like larva is much more difficult to see but can be found if you search carefully through ivy plants in spring.

How to find

▪ **Timing** This moth is on the wing from the end of June to the beginning of August.

▪ **Habitat** It occurs in gardens, woodland and more open, scrubby areas. The main larval foodplant is ivy, but it will also use other broadleaved plants.

▪ **Search tips** Swallow-tailed Moths have rather a brief flight season. Using a moth trap in early July gives the best chance of seeing one; it will also come to lit windows.

WATCHING TIPS

Ivy in the garden is not always popular but will encourage myriad insects, including this lovely moth. If you have time to spare and already know the species visits your garden, have a gentle look in ivy clumps for the caterpillar in spring. Its camouflage is quite remarkable – it moves with a looping action but if disturbed it will cling to a slim branch with its hind end and stick its head and body out straight to look like a broken twig.

Migrant lepidoptera

It is one of the marvels of nature that butterflies and moths, despite their small size and fragility, can and regularly do fly vast distances and even make significant sea crossings. Most species are, of course, highly sedentary, but among the migrants are some of the natural world's most impressive travellers. Britain's position means that the proportion of migratory species on our national list is much higher than for similar-sized but landlocked countries, and although some species are regular and predictable migrants, others wander here only occasionally and are perhaps better considered as vagrants.

Which species?

The best-known regular migrants are the Painted Lady and Clouded Yellow butterflies, and among moths the Silver Y and Hummingbird Hawkmoth. None of these can normally survive our winter, but come here

The **Short-tailed Blue** is an infrequent and unpredictable migrant visitor.

In a bumper **Painted Lady** year, butterflies will often be arriving in large numbers as early as May.

in sometimes huge numbers in summer – there is evidence of a return autumn migration in at least some cases. Witnessing a major movement of any of these species can be an astounding experience. One such was recently caught on camera by accident – the final of the Euro 2016 football tournament in Paris drew in huge numbers of migrating Silver Y moths, diverted from their northward course by the stadium lights.

Rare migrants include some of the most sought-after species. The Monarch, a stunning large butterfly native to North America, regularly turns up in south-west England, one of only a handful of insects that have successfully made the Atlantic crossing. Rare butterflies reaching us from mainland Europe include Short-tailed and Long-tailed Blues, Queen of Spain Fritillary, Swallowtails of the European subspecies, and Pale and Berger's Clouded Yellows. Notable moth migrants include Convolvulus and Death's Head Hawkmoth, and the glorious Clifden Nonpareil.

When do they arrive?

Migrations to Britain from continental Europe are usually triggered by high numbers in areas where the species is resident, alongside weather conditions that encourage movement north and/or west. The first factor will vary between species, which is why a big

Clouded Yellow year isn't necessarily a big Painted Lady year, but the right wind direction will definitely play a part for all migratory insects. Monarch arrivals usually happen following westerly gales, and almost invariably at the same time that numbers of North American birds also turn up on south-west coasts.

There are many wildlife-watching groups on Facebook and other social media which share information about migrating insects. Make sure you are part of these groups, for advance warning of arrivals.

How to find them

As with migratory birds, the best places to encounter migrant moths and butterflies are coastal headlands in the South and East (and South-West for Monarch and other North American species). Butterflies and day-flying moths may stop immediately to feed, or continue inland. Pick a headland with some sheltered, flowery dips that might tempt the insects to pause. You are unlikely to witness the arrival of nocturnal migrant moths but if you live or are on holiday on the coast, be sure to run your moth trap, especially after mild breezy nights with the wind coming from the near continent. All interesting sightings should be passed on to local butterfly and moth recorders – find yours here: butterfly-conservation. org/2390/Recording-contacts, and mothscount.org/text/57/ county_moth_recorders.

The **Silver Y** is our most frequent and conspicuous migratory moth, with plenty arriving in most years.

In Britain, most sightings of **Long-tailed Blue** occur on the south coast in autumn. More than 100 were found in 2013.

Red Underwing

Catocala nupta

Wingspan 7cm

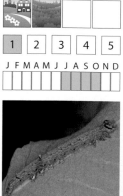

1	2	3	4	5

J	F	M	A	M	J	J	A	S	O	N	D

This is a large and very attractive moth that livens up the moth-watching season as autumn arrives. The forewings are not colourful but bear a beautiful, complex pattern of waves and scallops in shades of grey and brown. When startled, it reveals the hindwings which are shocking red, edged with black and with a black central band. It is one of a number of related but rare species with red underwings, and also the spectacular Clifden Nonpareil, a very rare but possibly increasing migrant which is even larger and has a royal blue band on its hindwing. However, the Yellow Underwing and its relatives belong to a different family.

Quite a number of moths make use of 'flash' coloration – if detected by a predator despite the camouflaged forewings, it rapidly opens them to show the bright colours of the hindwings, hopefully distracting the predator for long enough to make its escape.

How to find

■ **Timing** Adults are on the wing at night in August and September. They are nocturnal, though may also be found by day resting on walls or tree trunks, especially after a still, warm night.

■ **Habitat** This species is found in woodlands, parks, scrubland and other habitats where willow and poplar (the main larval foodplants) grow.

■ **Search tips** Its well-camouflaged forewings make this species hard to spot, despite its considerable size. It is most easily found by using a moth trap – it is also attracted to night-scented flowers.

WATCHING TIPS

As well as light, it is possible to attract some moths, including the Red Underwing, by 'sugaring'. This involves making up a pungent mixture of brown sugar dissolved in beer, with some very ripe fruit and/or dark syrup added if you have them handy. Paint the mixture onto tree trunks, stay up late with a torch, and see what arrives. The best nights for sugaring are warm with a light breeze, and ideally overcast rather than moonlit. This method gives you the chance to watch the moths' feeding behaviour as well as to see them close-up.

Emerald moths

Large Emerald moth

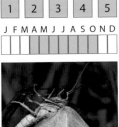

| 1 | 2 | 3 | 4 | 5 |

J F M A M J J A S O N D

Common Emerald moth

There are a number of species of emerald moths in Britain, all with shining light mint-green wings, which sadly fade rather quickly after emergence. There is usually a double white stripe on the forewing, and one or two on the hindwing. Some are common, others very localised – most species are commonest in southern England. They rest with wings fully spread. The species most likely to be encountered are Common, Little and Large Emeralds. The most striking is the Blotched Emerald, which has scalloped cream and black wing margins and some pale creamy patches on the forewing. The Blotched Emerald caterpillar improves its camouflage further by attaching pieces of leaf to its body.

Small Emerald moth

How to find

▪ **Timing** Emerald moths are nocturnal, and most species fly in mid-summer.
▪ **Habitat** Most of these moths are commonest in wooded areas, as their foodplants are various species of trees and bushes.
▪ **Search tips** Emerald moths are attracted to light. They are well-camouflaged in foliage but stand out if they happen to rest on walls or tree trunks, so are among the most frequently spotted nocturnal moths.

WATCHING TIPS

These moths are attractive visitors to many garden moth traps. They also have interesting traits in the larval stage, including a colour change from brown in autumn to green in spring in some cases. Of the more localised species, look out for the bright turquoise-green Grass Emerald on moorland in Scotland. It is always worth looking out for emeralds at the very start of their flight season, while their remarkable colour is at its best.

Pugs and carpets

Common Pug moth

1 2 3 4 5

J F M A M J J A S O N D

Silver-ground Carpet moth

Common Carpet moth

The family Geometridae is very large and a substantial number of its species are the small species known as pugs and carpets – mostly placed in the subfamily Larentiinae. Carpets are typically small moths with a complex 'wave' pattern across the wings, while pugs are even smaller and have long, narrow wings, the wingspan far exceeding the body length. All have slim and delicate bodies. Pugs mostly rest with wings fully spread, while carpets more often hold the forewings covering the hindwings, making a triangular shape. There are many species in Britain, some common, some extremely rare, and many representing a real identification challenge. In a few cases, females are wingless and use pheromones to attract a mate.

How to find

■ **Timing** Most species are active by night between May and August, but there are at least a handful of species on the wing in each month of the year – for example, the Winter Moth flies in January and February.

■ **Habitat** Pugs and carpets span all kinds of habitats between them. None are powerful fliers so their distribution is limited by the presence of larval foodplants. Often the clue is in the name, for example Foxglove Pug, Toadflax Pug and Spruce Carpet.

■ **Search tips** These little moths are among the species that you might see if you go on an organised 'bug hunt' at a nature reserve, and use a sweep net to catch insects disturbed from foliage. Many are also attracted to light – if you use an outside light, check the adjacent walls and fences first thing in the morning.

WATCHING TIPS

Many pugs and carpets are small enough to be classed as 'micro-moths' – an imprecise definition covering all species with a wingspan of less than 2cm. In general, they are less well-known and less studied than the macro-moths and there is much for amateur lepidopterists to contribute here. Certain species of carpet are also exceptionally pretty and well worth the effort to locate – among the real beauties are the Flame Carpet, Red-green Carpet, and the aptly named Beautiful Carpet.

Other geometrids

1	2	3	4	5

J F M A M J J A S O N D

Brimstone moth

Blood-vein moth

A large proportion of British moths are geometrids. They include familiar species such as the bright yellow Brimstone Moth, the red-striped Blood-vein, the Scorched Wing with its dusky wingtips and curled abdomen, and the striking Canary-shouldered Thorn with its fluffy, bright yellow body. Some geometrids have two or more distinct colour morphs, most famously the Peppered Moth, which played a key role in our understanding of natural selection through the phenomenon of industrial melanism. Most geometrids will come to light – a few are day-flying. The greatest variety of species is found in southern England, but several are found only in other parts of the British Isles.

Peppered moth

How to find

■ **Timing** More species fly between June and July than at other times, but there are still plenty of species on the wing in spring and again in autumn.

■ **Habitat** All kinds of habitats support different geometrid moths. There are likely to be dozens in your garden or local park.

■ **Search tips** Using a moth trap in the garden is a great first step to finding out which geometrids are present in your local area – so is attending a moth-trapping night at your local wildlife reserve. If you are out and about early in the morning in summer, check lit shop windows and adjacent walls for resting moths. Most modern phones are capable of taking good macro images, so you can identify what you see later on.

WATCHING TIPS

Moth-watching is both addictive and challenging. The moth trap is the best way to meet your local species, and many moth-watchers set the trap at the same time each night through spring to autumn, to monitor which species emerge when, and how their numbers rise and fall through their flight season. If you prefer a more ad hoc approach, though, you can try taking the trap to other locations. The geometrids are a diverse group and studying them will be a substantial and rewarding project.

Silver Y

Autographa gamma

Wingspan 3.7cm

| 1 | 2 | 3 | 4 | 5 |

J F M A M J J A S O N D

A dusky, mottled grey moth of the large family Noctuidae (see p162), this species is named for the white marking on its forewing which resembles the letter y, or the Greek character **Y**. It has raised ridges of hair on its thorax and where its wings meet, forming a series of crests. This is the most easily observed migrant moth in Britain, and can be seen anywhere in the British Isles – in considerable numbers in some years. Those that arrive will often breed, but they are not usually able to survive our winters. There are several similar species that are resident in Britain, such as the Beautiful Golden Y, but the Silver Y is active by day as well as night and so is more likely to be seen than its relatives.

The first wave of arriving Silver Ys will breed in Britain and Ireland and produce a home-grown generation of adults, which are joined by yet more immigrants as summer progresses. In some years there will be vast arrivals and you will find them almost everywhere.

How to find

■ **Timing** Silver Ys start to arrive in mid-spring, and continue to turn up through summer and autumn, while early arrivals produce a home-grown generation to increase numbers further. Their numbers are highest in mid-summer. Silver Ys will feed in daylight as well as at night.

■ **Habitat** This moth appears wherever there are suitable flowers for nectar. It is often encountered in meadows, on downland, woodland edges, waste ground and parks. It lays its eggs on Common Nettles and a range of other herbaceous plants.

■ **Search tips** A fast-buzzing moth visiting flowers in the daytime is likely to be this species. Look for it anywhere in the countryside, but particularly near coasts and on south-facing hillsides.

WATCHING TIPS

These moths feed from flowers without fully settling, and their constantly buzzing wings are eye-catching – this can also make identification difficult so be patient. Numbers can be really impressive in good years, and a large influx of Silver Ys means it is worth looking out for scarcer migrants as well, as it is a sure sign that conditions have been good for making the sea crossing. The Silver Y migrates mainly at night and, trying to navigate by moonlight, is strongly attracted to light.

Griposia aprilina

Merveille du Jour

Wingspan 4.6cm

1	2	3	4	5

J F M A M J J A S O N D

This moth's lovely colours fade to whitish within days of emergence.

This moth's name translates as 'marvel of the day', and while it is actually nocturnal it is certainly a marvel – one of our most beautiful noctuid moths and much coveted by moth-watchers. The wings are pale soft green, overlaid with an intricate lace-like black-and-white pattern. When seen side-on, it shows a small crest of hair on its thorax. Its colours and pattern become faded rather quickly after emergence. It occurs widely but rather patchily in Britain and Ireland, becoming scarcer in north Scotland. Its much rarer relative, the Scarce Merveille du Jour of southern England, is very similar but with heavier black markings.

How to find

■ **Timing** This is a late-season species, on the wing in September and October, and active at night.

■ **Habitat** It is found in parks, woodlands and sometimes large gardens. The larval foodplants are oaks.

■ **Search tips** If you live in an area with plenty of oaks, look out for this species in the woods (its wings provide camouflage against lichen-mottled bark) and set your moth trap on autumn nights.

WATCHING TIPS

This lovely moth is difficult to observe but with luck and patience you could see it visiting flowers, or find it by day tucked inside a bark crevice on an oak trunk. The beautifully marked caterpillar is nearly as impressive a find as the adult moth, its colours and patterns echoing those of its adult form. Look for it feeding on oak flower buds in spring, and on leaves later in the year. The Scarce Merveille du Jour also uses oaks as larval foodplants, but its lifecycle runs to a different schedule, with adults flying earlier in the year (June and July).

Noctuid moths

Xyleninae and Hadeninae

1	2	3	4	5

J F M A M J J A S O N D

Coronet moth

Angle Shades

Noctuidae is another large moth family. Most have brown or grey wings, are quite plump in the body, and resting with wings drawn well back to create a narrow triangle shape. Many noctuids still bear the charming common names given to them by Victorian entomologists – within the large subfamily Xyleninae are such gems as The Uncertain, The Rustic, and The Confused. Hadeninae is another large family, which includes the wainscots and brocades among many others, and the least inspiringly named moth of all – Lead-coloured Drab. Most are nocturnal and drawn to light; they include familiar species such as the Angle Shades, and rarities the beautiful Bordered Gothic.

Most noctuids are quite robust, and while they often have intricate patterning and unusual wing shapes, only a few species sport any bright colour to speak of. Their excellent camouflage makes them difficult to find, but they will come to light.

How to find

■ **Timing** Xyleninae and Hadeninae noctuids of various species can be found at most times of year, though June to August is the peak period. Nearly all species are nocturnal.
■ **Habitat** These moths are found in all kinds of habitats, with woodland and lush open countryside the most productive hunting ground. Many visit gardens.
■ **Search tips** The best way to find these moths is, as ever, to use a moth trap. Some will also come to sugar. To seek out scarce and localised species, join your local moth group, who will be able to give advice. When on a leisurely country walk, check shaded, lichen-marbled walls and tree trunks for resting moths – they are very easily overlooked, but practice will greatly improve your moth-finding skills.

WATCHING TIPS

This group of moths, while not as showy as some, include some very attractively marked species. When you empty the moth trap, it's worth taking the time to take clear photos of the freshest specimens against the same plain background, to create your own reference library. Also look out for caterpillars on your garden plants – as you would expect, native plant species are far more likely to support caterpillars than exotics are.

Noctuid moths continued

Beautiful Golden Y moth

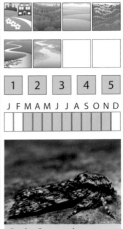

1	2	3	4	5

J F M A M J J A S O N D

Poplar Grey moth

While Xyleninae and Hadeninae are the largest noctuid subfamilies in Britain, there are about 16 other smaller subfamilies, of which Noctuinae is the largest. This group includes the Large Yellow Underwing, probably the most frequently encountered noctuid of all. Noctuinae moths tend to be rather plain, often with just one or two distinct wing markings – this is reflected in names like Heart and Dart, and Small Square-spot. However, some are stunning, such as the intricately marked True Lover's Knot. The smaller subfamilies include Plusiinae, with the gorgeously iridescent Burnished Brass its most spectacular species, and Acronictinae which includes the dagger moths.

Finding adult noctuid moths as they rest by day is an art that you can improve with patient practice. Most seem to rest instinctively against something that offers good camouflage. This is their only defence against daytime predators such as birds.

How to find

■ **Timing** This group includes some conspicuous day-flying species, such as the delightful Small Yellow Underwing. The majority are nocturnal though, and most are on the wing between May and August. Some species have conspicuous caterpillars that you'll find a few weeks prior to or after the flight season.

■ **Habitat** These moths occupy all kinds of habitats, and some will breed in gardens. For example, caterpillars of The Mullein are often found on buddleia.

■ **Search tips** Most of these moths are attracted to light, so using a trap is the best way to find them. Your local moth group will be able to help you locate particular species. Nature reserves often hold moth-trapping nights, and/ or summer fairs in which moths trapped the night before will be on view.

WATCHING TIPS

This is a large and confusing group of moths, and their English names do not necessarily help you to work out which are closely related. Discovering which species use your garden or local wildlife-watching patch is the natural starting point, but it will also help enormously to get to know other moth enthusiasts, who will be able to show you new species and provide hints on how to get the best from your trap. Do take the time to place moths in cover, out of sight of predators, after you have checked and photographed them.

Longhorn moths Adelidae

Wingspan up to 2cm

Yellow-barred Longhorn

1	2	3	4	5

J F M A M J J A S O N D

Green Longhorn

Although they are very small, the longhorn or 'fairy longhorn' moths are very noticeable because of their showy behaviour, the males 'dancing' in groups in sunlight on spring and summer days. The name relates to their unusually long fine antennae – in some species the male has antennae many times longer than his body, usually with white tips that catch the light during the 'dance'. The wings often have a metallic gloss, in green, purple or red tones. This is a small family, and most species do not have common names. The two most often observed species are the widespread *Nemophora degeerella* (Yellow-barred Longhorn) and *Adela reaumurella* (Green Longhorn).

The small body size and preposterous antenna length makes these species rather a challenge to photograph! Those photographers who really want to test their skill could try capturing the males in their up-and-down aerial dance.

How to find

■ **Timing** Most longhorns are on the wing between May and July. They are active on warm, sunny days without too much breeze.

■ **Habitat** These moths occur in sheltered lush habitats, such as meadows with some scrub, woodland glades and edges, and hedgerows.

■ **Search tips** The swarming males are eye-catching. Search for them around leafy shrubs such as hawthorn, in full sun and sheltered from the wind. They may give the impression of a swarm of flies at first glance, but they land frequently, giving clear views.

WATCHING TIPS

Longhorn moths are delightful to watch, their absurdly long antennae whipping about in all directions as they bob up and down. Females ready to mate will approach the swarm and are pounced on by the nearest male – this happens very quickly but watch for a while and you could observe it. Some longhorns have gorgeous coloration, in particular the rare *Nemophora fasciella* with its violet and copper wings, and are worth the effort to photograph.

Swift moths

Ghost moth (female)

Wingspan 3.3–4.5cm

1	2	3	4	5

J F M A M J J A S O N D

Ghost moth (male)

This family has just five British representatives, though world-wide there are hundreds more. They are anatomically distinct from other moths in several ways and are more primitive than other macro-moths. One particular trait they share is in having distinctly different-looking males and females – in most of our other moths the sexes are alike or nearly so in appearance, though females are usually a little larger. Swift moths have fluffy thoraxes, very short antennae, and hold their forewings in a tent shape. Most species are brown with some paler or white markings – the Map-winged Swift has particularly attractive markings, while the male Ghost Moth is pure white. All are fairly widespread through Britain and Ireland.

These moths have a distinctive sleek outline, and their short antennae are usually held back against the head so are difficult to see. The male Ghost Swift with his pure white wings is a striking sight on summer nights.

How to find

■ **Timing** Swift moths mainly fly in late spring and summer, with the Orange Swift having a later season (July to September). They are nocturnal but often begin to fly a little before dusk.

■ **Habitat** Swift moths mainly occur in open habitats, such as moorland and hilly meadows. The larvae are rarely seen as they feed on the roots of various grasses and other plants (and may take two years to fully develop).

■ **Search tips** These moths will come to light. They do not feed as adults and so males very actively search for females through their few days of life. This means that they are often found away from their usual habitat.

WATCHING TIPS

Swift moths epitomise the 'live fast, die young' adage, both sexes being very active. Females do not even take the time to land when laying eggs but simply scatter them in flight. This means that if you trap a female, she may well lay numerous eggs in the trap. Unfortunately, the larvae's subterranean feeding habits and long maturation time make them near impossible to rear in controlled conditions. Occasionally, Common Swift and Ghost Swift larvae can do noticeable damage to herbaceous plants in the garden.

Prominent moths and relatives

Iron Prominent moth

1	2	3	4	5

J F M A M J J A S O N D

Puss Moth caterpillar

The family Notodontidae is mainly represented in Britain by the 'prominents', which tend to have prominent ridges of hairs on their thoraxes, giving a distinct humped shape. They also have noticeably fluffy legs, especially the front pair. Related species include the Lobster Moth, a large grey, very furry moth named for its extraordinary pink prawn-like caterpillar. Another bizarre caterpillar is that of the Puss Moth, a squat green creature which, when mature, bears a large startling 'false face' on its expanded front end, and long double 'tails' at the other end. Another well-known member of the family is the Buff-tip, which perfectly resembles a short, broken piece of birch twig.

This moth family boasts some of the most diverse and peculiar-looking caterpillars you are ever likely to find anywhere. The Puss Moth and related kitten moths have quite monstrous-looking larvae, but are destined to become very pretty and fluffy-looking moths.

How to find

▪ **Timing** These moths are mainly on the wing in summer, though some are double-brooded so may be seen from spring through to autumn. They are nocturnal.

▪ **Habitat** Most are woodland species, their larvae feeding on trees of various kinds.

▪ **Search tips** Many of these moths are attracted to light. They can also be found in the open if you look carefully, as they have camouflage intended not so much to conceal but to resemble some scrap of lifeless organic matter: a dead leaf, piece of twig or a bird dropping. If disturbed, they take the mimicry a stage further by dropping motionless to the ground rather than flying away.

WATCHING TIPS

The prominents and other notodontids are all very attractive and fascinating moths, often as much so in their larval stage as when adult. Larvae often have a special 'camouflage' posture, while the remarkable Puss Moth caterpillar is more assertive and shows its false face and raises its tail projections menacingly if threatened. Be particularly careful when handling the adults during release from the moth trap as their habit of playing dead makes it easy to drop and lose them.

Pterophoridae

Plume moths

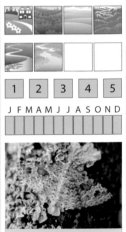

1 2 3 4 5

J F M A M J J A S O N D

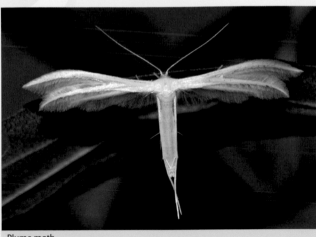

Plume moth

Twenty-plume moth

The plume moths are quite striking, although not immediately recognisable as moths – or even as living things – at first glance. The wings are divided into separate, feathery branches, but at rest are rolled up into a narrow tube and usually held roughly at right angles to the narrow, slightly bulbous-tipped body. A resting plume moth is therefore shaped much like a capital T, with long, thin antennae and spindly legs attached. There are dozens of species known in Britain, most with no common names. The Twenty-plume Moth is in a different family but has the same feathered wing form – however, it holds its plumes spread out rather than rolled up. Plume moths often come to lit windows and may turn up in moth traps.

The 'true' plume moths (and the enchanting Twenty-plume Moth as well) are all creatures of delicate beauty, well worth examining with a hand lens. They form a confusing group, but some species can be identified with reasonable confidence as long as you have a good look.

How to find

▪ **Timing** Most species are active in spring and summer but some are double-brooded and fly into autumn and even winter, hibernating through the coldest weeks in their adult form. The Twenty-plume Moth breeds continuously and can be found at any time of year.

▪ **Habitat** Many species occur in a wide range of habitat types, while a few are specialists – *Platyptilia isodactylus*, for example, occurs in wetlands, and *Agdistis bennetii* on saltmarsh. Their larvae feed on various herbaceous plants.

▪ **Search tips** The commoner plume moths are often discovered resting on walls, fences or windows close to outside lights. When in more natural surroundings, they tend not to conceal themselves but are easily mistaken for bits of dried grass, so search slowly and patiently to find them. The larvae are well-camouflaged on their foodplants and hard to find.

WATCHING TIPS

Plume moths are not very active, though adults will come to nectar sources at night. Identification can be extremely difficult and in some cases definite identification is only possible through genital dissection of specimens – which may be a step too far for many wildlife-watchers! However, amateur observations of all plume moth species are very helpful in continuing to build knowledge about this diverse and relatively little-known group.

Hooktips and related species

Drepanidae

Oak Hooktip

1	2	3	4	5

J F M A M J J A S O N D

Peach Blossom

In Britain there are rather few members of the family Drepanidae, but they include some very distinctive moths. The hook-tips are so called because the forewings have curved and pointed tips, while the other main group in the family, the lutestrings, have dark lines running horizontally across their wings. The group also includes one of Britain's most beautiful moths, the widespread Peach Blossom, named for the splashes of rose-pink on its dark brown wings. It is closely related to another very attractive moth, the Buff Arches, which has a complex red-brown and white pattern but also distinctive large patches of smooth unmarked grey. Most drepanid larvae are strongly patterned and sometimes peculiarly shaped, and often rest with head and tapered tail raised.

The family Drepanidae is mainly found in eastern Asia, with relatively few species occuring in Europe and only 15 or so in the British Isles. Most of them are distinctive and quite easy to identify.

How to find

■ **Timing** Most are summer-flying species, and they are nocturnal. Some produce two broods a year.
■ **Habitat** Most of the drepanids lay their eggs on one or a few species of deciduous tree or shrub, and so are most likely to be found in deciduous woodlands or woodland edges.
■ **Search tips** Drepanids will come to light, and to sugar. They otherwise spend most of their time up in the trees, and so are difficult to find.

WATCHING TIPS

These interesting moths are quite a challenge to observe in the wild, but the average garden moth trap stands a good chance of attracting at least a couple of species, including the Peach Blossom. This moth lays its eggs on bramble, so allow a clump of bramble to flourish in your garden if possible – its flowers and fruit will attract plenty of other interesting insect life too. If you live in south Wales or south-west England and there are small-leaved lime trees in your area, they could support the Scarce Hooktip, one of the rarest and most attractive drepanids.

Micro-moths

Small Magpie moth

Wingspan up to 2cm

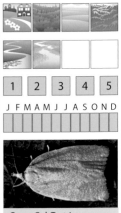

| 1 | 2 | 3 | 4 | 5 |

J F M A M J J A S O N D

Green Oak Tortrix

There are nearly 2,500 moth species recorded in Britain, of which most are 'micro-moths' – those with a wingspan of 2cm or less (some larger species are also considered to be micro-moths because their close relatives fit the 'micro' definition). The distinction is not scientific, but has nonetheless been used as a basis for separating moths into two groups. The micro group has frequently been left out of field guides entirely because so many are incredibly difficult to find and identify. Interest in micro-moths is now growing, though they are still a group for the specialist. However, some species are remarkably beautiful, interesting, and worth the trouble to find.

Small can be beautiful – seen through a hand lens or a macro lens, even the most miniscule micro-moth may prove as stunning as any big, showy butterfly, with their own array of bright colours and pretty patterns.

How to find

■ **Timing** Micro-moths of various species fly at all times of year, and some are diurnal. The greatest diversity of species is active on summer nights, but the variety is less overwhelming at other times.

■ **Habitat** These moths are found in all habitat types – many will inhabit even the smallest garden, and a few are found almost exclusively inside homes and other buildings.

■ **Search tips** Whenever you are out in the countryside you will find micro-moths, or at least evidence of their activities. Scouring tree bark and the undersides of leaves may produce adult moths. Also look out for 'leaf mines' – discolouration in distinct patterns on leaves, indicating that a tiny micro-moth larva has been feeding on the tissues within. Some mines are long squiggly tunnels that start narrow and widen as the larva grows, while others are round blotches. Different species produce different mine types.

WATCHING TIPS

Getting to know Britain's great variety of micro-moths is a task that will take several lifetimes, but you can start by seeking out some of the more distinctive species. One such is the stunning *Alabonia geoffrella*, which is active by day, and has vivid metallic blue markings on golden and white wings. The Diamond-back Moth draws attention by its numbers rather than appearance – it is a noted migrant and huge influxes occur along the coast from time to time. Tortrix moths are often noticed in their larval stage – caterpillars fall from the trees on a thread of silk when disturbed.

Day-flying moths

If you are daunted by the sheer number of moths present in Britain, you can make life considerably easier by concentrating on those that fly by day. Not only are they relatively few in number, but they are much easier to observe in a natural state. Also, they include some particularly attractive species, with representatives from many different families. Most day-flying moths are found in open countryside rather than in woodland, and behave rather like butterflies, flying in full view and stopping to feed from flowers.

Species to look for

Several of the more striking and widespread day-flying moths have been covered in detail in previous pages, such as the Hummingbird Hawkmoth, the burnets, the Silver Y and the longhorn moths. Here are some of the others, and when and where to find them.

Chimney Sweeper moth

Chimney Sweeper (above) A distinctive small black geometrid moth with narrow white wingtips and a fluttering, butterfly-like flight. Look for it throughout Britain in damp and dry grassland, and moorland edges, in June and July.

Burnet Companion (left) This moth, of the family Erebidae, appears in late spring, a little before most burnets are on the wing, and is often found on downland and in sheltered meadows. It has very velvety-looking wings with a subtle but pretty pattern in brown and grey, and the hindwings are mostly yellow. It is most common in the South.

Burnet Companion moth

Mother Shipton (left) Related to the Burnet Companion and similar to it, this moth's forewing pattern resembles a witch's face in profile. It is a widespread species of flowery meadowland and flies in May and June.

Small Purple-barred This is a colourful erebid micro-moth, with bright purple-red stripes across mossy yellow-green wings. It is found from May to June in grassland and heathland.

Mint Moth (top opposite) Flying through summer in two generations, this micro-moth of the family Crambidae is common in gardens. It has purple forewings and dark hindwings, each marked with yellow spots.

Mother Shipton moth

Mint moth

Argent and Sable A small geometrid, this moth is uncommon – most likely to be seen in Scotland. It has white wings with extensive black blotches, and flies on summer days in boggy moorland and woodland edges.

Latticed Heath (below) This is a common moth of heaths, grassland and other open countryside, flying from spring into autumn. Its pale wings bear a pattern of evenly spaced black stripes and bars.

Speckled Yellow (below) Another small but broad-winged geometrid, the Speckled Yellow flies in May and June in open scrubby areas. It has yellow wings marked with brown, black-edged spots.

Beautiful Yellow Underwing One of the few diurnal noctuids, this small moth is indeed a beauty with its red-and-white marbled forewings and yellow, black-bordered hindwings. It is found patchily though Britain on heather moorland, flying from May to August (June and July in Scotland).

Latticed Heath moth

Speckled Yellow moth

Emperor Dragonfly

Anax imperator

Length 6cm

| 1 | 2 | 3 | 4 | 5 |

J F M A M J J A S O N D

This very large, colourful dragonfly is widespread and common in England and south Wales, and is pushing north towards Scotland – it is also present in southern Ireland. The male has a green head (including eyes) and thorax, and a blue abdomen with a dark line down the centre. The female is similar but her abdomen is green (there is also a rare blue female form). This dragonfly is more solidly coloured than the blue and green hawker species, which have distinct coloured spots on a black background. The wings show a slight yellow tint, and the abdomen tip often droops downwards.

Male Emperors pause quite often to rest on waterside vegetation, before resuming their patrol in search of females. When a female appears, the pair mate in a quick mid-air scramble then go their separate ways.

How to find

■ **Timing** Emperors first appear in late May or early June, and fly until mid-August. They are most active in the warmer hours of the day.

■ **Habitat** This species is rarely seen away from open water with well-vegetated margins. It can be seen over lakes and even small ponds, as well as canals and slow-flowing rivers, and marshy lagoons.

■ **Search tips** Male Emperors are very eye-catching as they tirelessly patrol the area of water that they establish as a territory, hunting on the wing while looking out for females. Look out for female Emperors perched on floating debris, ovipositing into the water.

WATCHING TIPS

This is a dramatic dragonfly and one of the first large species to appear each year. Territory-holding males fly for long periods but regularly pause on low perches overhanging the water, and settle if they catch large prey. Encounters between the sexes are brief mid-air affairs. After mating you may see the male curling his abdomen tip up under his body, to pass fresh sperm from his primary to secondary genitalia. Ovipositing females often attract the attention of damselflies, which appear to mob them, though the real reason for the behaviour is unknown.

Aeshna cyanea

Southern Hawker

Length 6cm

1	2	3	4	5

J F M A M J J A S O N D

Female Southern Hawkers with eggs to lay can be easy to watch. They usually settle at the water's edge, unlike female Emperors which prefer to land on floating vegetation when egg-laying.

This dragonfly is similar in size and colour to an Emperor, but its markings are spots on a black background. The male has blue eyes and blue bands on the tail-tip and on the abdomen underside – the markings are otherwise apple-green. The female has duller eyes and green rather than blue on the tail-tip – she also lacks the male's slim 'waist'. The best field marks are the pair of large round green spots on top of the thorax, and the last three tail-tip markings, which join up as continuous bands. This hawker occurs throughout England and most of Wales, and is found at some coastal parts of north Scotland – it is also present (though rare) in south-west Ireland.

How to find

■ **Timing** Southern Hawkers typically appear in July, and are on the wing well into September and beyond if the weather is mild enough.

■ **Habitat** Southern Hawkers are often found well away from water, for example on open heath or woodland edges. When seeking to mate, they frequent lakes, ponds, slow-flowing rivers and other lush water bodies – they will even visit small garden ponds.

■ **Search tips** This is a restless hawker, flying fast and often in straight lines along pathways or hedgerows quite low to the ground – sometimes you will hear the rattle of its wings before you see it. You may find it 'hanging up' on a perch when the weather is cooler, but it is hard to spot when not moving. It will continue to hunt late into the evening. Females with eggs to lay stick to the margins of water bodies.

WATCHING TIPS

This species shows real interest in human observers, often hovering very close to your face. Its fearlessness makes it an easy species to watch, and its agility in the air is breathtaking – flying along at knee level, then shooting vertically upwards to the treetops to chase down prey. Small prey is eaten in flight but if it captures something substantial, such as a butterfly or another dragonfly, it will land to subdue and consume it. Females are well known for trying to oviposit in odd places – anywhere low down and close to water will do, even a wildlife-watcher's boot.

Migrant Hawker

Aeshna mixta

Length 5cm

1	2	3	4	5

J F M A M J J A S O N D

This hawker is not quite as hyperactive as some of its cousins, and males will often pause for a short rest on waterside vegetation before resuming its patrols. Sometimes several will gather to attack swarms of midges or mosquitoes.

Once a rare migrant visitor to Britain, this species is now an established resident, occurring widely across England, Wales and southern Ireland, and spreading north. It can be very abundant at good sites in September. It is a little smaller than Southern and Common Hawkers, and unlike these species the male does not have a slim 'waist'. The male has mostly blue eyes, a dark thorax with yellow stripes on the side, and paired blue spots down the top of the abdomen, with a yellow 'golf tee' marking at the very top of the abdomen. The double pale markings on the top of its thorax are just tiny dots. The female has yellow-green rather than blue spots, smaller than the male's spots, and greyish eyes.

How to find

■ **Timing** The first Migrant Hawkers appear in late July, with migrants from the continent soon joining the home-bred population. September is the peak month, and a few may hang on into November.
■ **Habitat** Any kind of wetland can support this dragonfly, though large systems of lakes and ditches hold the largest numbers. Immature individuals may also be seen in open habitats away from water.
■ **Search tips** Because this dragonfly is not very territorial, you may see several using the same water body, or attacking the same swarm of midges – it will even roost in small groups. Males are very visible as they fly along the edges of lakes or canals, sticking to a particular patch but tolerating intrusions from other males. They often pause to hover on the spot for a few seconds at a time.

WATCHING TIPS

Migrant Hawkers are easy to watch and, with their hovering habits, much easier to photograph in flight than most dragonflies. Patrolling males are quick to seize females that approach the water, and once paired the two will often settle and remain perched for a long time. Unaccompanied females are not seen as often but you may spot them ovipositing at or near the water's edge. This species is on the wing longer into autumn than any other large dragonfly species – it and the Common Darter liven up many a sunny late-autumn visit to wetland areas.

Aeshna juncea

Common Hawker

Length 5.8cm

1	2	3	4	5

J F M A M J J A S O N D

This species is sometimes known as the Moorland Hawker, a more apt name as, although widespread, it is rarely common. Its range has a northern and western bias, with the strongest populations in upland Scotland, Wales, north Ireland and the south-west of England. The male has blue eyes and abdomen spots, and yellow thorax stripes, including two narrow stripes on top of the thorax. Its blue markings are noticeably darker in colder conditions. Females have smaller spots, which are typically yellow-green but may be green or occasionally blue – they lack the male's pinched-in waist at the top of the abdomen. Both sexes show a striking yellow costa (the vein at the leading edge of the wing).

Like many other dragonflies and damselflies, the Common Hawker's colours are affected by air temperature and it can look very black when it has not had a chance to warm up. It will often choose rocks facing away from the wind as basking spots.

How to find

■ **Timing** Common Hawkers fly from July through to mid-autumn, into November.

■ **Habitat** It is most reliably found on open heaths and upland heather moor with some reasonably large pools, but will also breed at lowland waters in woodland, and will hunt around woodland well away from water.

■ **Search tips** This is a fast-flying and often nervous hawker – easy enough to spot on the wing in its powerful flight, but difficult to approach when it settles. You may have more luck on a day with mixed cloud and sunshine, as cooler conditions discourage flight. It enjoys basking, so you may have success if you wait by a large sun-warmed rock or patch of bare ground. Listen for females rattling their wings as they stand on floating vegetation, laying their eggs.

WATCHING TIPS

Views of this hawker are often fleeting, and you will need luck or patience to see one close-up, but binoculars are helpful to watch behaviour from a distance. Males patrol the water's edge, fiercely chasing away other males and grabbing any females they encounter. If the female is receptive, the pair move to low vegetation to copulate and may stay put for up to an hour. Scan ahead with binoculars when searching and you might spot one warming up on a rock on the path ahead.

Azure Hawker

Aeshna caerulea

Length 4.8cm

1	2	3	**4**	5

J F M A M J J A S O N D

This is a fairly small, rare hawker with an early flight season. It is found only in west Scotland, especially in the Wester Ross area, where its range overlaps with that of the Common Hawker. The male Azure Hawker differs from the male Common Hawker in having larger blue markings, and no yellow markings at all – its thorax stripes are blue. It is also smaller and lacks a slim waist. The female Azure has yellow-brown markings (occasionally blue). Both sexes have clear wings, without the yellow costa of the Common Hawker.

This beautiful insect is much sought-after by dragon-watchers.

Super sites

* Moorland alongside Loch Maree, Scottish Highlands
* Bridge of Grudie by Loch Maree, Scottish Highlands
* Glen Affric, Scottish Highlands
* Silver Flowes, Ayrshire

How to find

■ **Timing** Azure Hawkers appear at the end of May, and their flight season is over by early August.

■ **Habitat** This dragonfly is mainly found in heather moorland, using small shallow pools (tarns) for breeding. It will also hunt along streams and woodland edges.

■ **Search tips** This species is very much sought-after by dragonfly-watchers, and the fascination appears to be mutual. Azure Hawkers will readily approach human observers and even land on them – wear pale clothing to improve the chances of an encounter like this. Males explore all the pools in a small area in their search for females, and both sexes may be seen basking at the water's edge or on sun-warmed rocks. Azure Hawkers will not fly in overcast weather, so choose a day with at least some sunshine forecast, and allow plenty of time.

WATCHING TIPS

With patience it is not difficult to enjoy excellent close views of this dragonfly – the unreliable west Highlands weather is often the deciding factor. It settles much more often than Common Hawker and is much more approachable, flying close for a good look at you and, if you are very lucky, perhaps even landing on you. Watch for males searching the pools for potential mates, and coupled pairs resting on rocks or low vegetation close to the water.

Norfolk Hawker

Length 5cm

| 1 | 2 | 3 | **4** | 5 |

J F M A M J J A S O N D

This dragonfly's bright apple-green eyes are its most striking feature, distinguishing it from other mostly brown species.

This is the least colourful of our hawkers and is sometimes called Green-eyed Hawker – a more appropriate name, as in Britain it is no longer confined to Norfolk, and indeed it is very widespread in mainland Europe. The sexes are nearly identical, both being plain mossy-brown with yellow stripes on the thorax sides, a triangular yellow marking at the top of the abdomen, and bright green eyes. The species was formerly found only on the Norfolk Broads and similar habitats in Suffolk, but has spread south in recent years and is now present in north Kent.

Super sites

★ RSPB Strumpshaw Fen, Norfolk
★ RSPB Minsmere, Suffolk
★ Stour Valley, Kent

How to find

■ **Timing** This hawker emerges in June. Its flight season is very short, and few adults survive beyond the first half of July.

■ **Habitat** The Norfolk Hawker is most often seen along weedy ditches and canals with clean water, also slow-flowing rivers, and is associated with the wetland plant Water Soldier.

■ **Search tips** You are most likely to spot a male Norfolk Hawker, as he patrols his stretch of water in a low flight, looking for females and chasing off rival males. The species also perches more often than most hawkers, usually on reed or sedge stems at the water's edge. Keep an eye on clumps of Water Soldier as females seem invariably to lay their eggs among its leaves and stems (even though the larvae are, like all dragonfly larvae, strictly predatory).

WATCHING TIPS

This is quite an easy hawker to watch – not overly alarmed by human presence, and sometimes hovering in flight long enough for photographs to be taken. Its continuing spread, and the extent of its reliance on Water Soldier plants (which does not apply to continental populations) are both factors that the amateur entomologist can help to investigate. Males are very active around the middle of the day, patrolling a short stretch of river or ditch, and will fly in rather overcast days as long as it is warm enough.

Brown Hawker

Aeshna grandis

Length 5.8cm

1	2	3	4	5

J F M A M J J A S O N D

Brown Hawkers seem able to stay airborne for hours on end, but watch a patrolling hawker carefully and sooner or later it will land – hopefully not too high up for you to get a good look at it.

This is one of the easiest dragonflies to identify even from below-par flight views, because of its strongly golden-tinted wings. The body is dark brown, with sparse blue markings in the male – he also has partly blue eyes and a slim pinched-in waist, and both sexes have two yellow stripes on the thorax sides. This is a powerful large dragonfly with a confident level flight – it can even give the impression of a small bird at first glance. The Brown Hawker is widespread in England, eastern Wales, and in Ireland. The distribution thins out in the North and West, but just reaches south-east Scotland.

How to find

■ **Timing** This hawker has a long flight season, appearing in late June and flying to September or early October. It is very active and hard to approach in the heat of the day.

■ **Habitat** It is typically found in sheltered lush habitats with large lakes or slow-flowing rivers, and often hunts in woodland.

■ **Search tips** An eye-catching dragonfly, this species is often spotted flying several metres high, skimming tree foliage. You may also spot the females coming to the water's edge to lay eggs. Finding a Brown Hawker at rest is a rare treat – search carefully in the morning, looking quite high in trees and bushes (binoculars may help). They tend to be very flighty.

WATCHING TIPS

Watching a Brown Hawker tearing around high overhead gives a real insight into the predatory power of large dragonflies. Seeing one clearly at close range is rather more difficult but if you watch a hunting Brown Hawker long enough it will eventually settle. Males patrolling water in search of females are easier to watch as they fly lower down. Females often perch on floating logs to lay their eggs, letting their abdomens droop down into the water.

Brachytron pratense

Hairy Dragonfly

Length 4.5cm

1	2	3	4	5

J	F	M	A	M	J	J	A	S	O	N	D

This hawker has a similar colour scheme to Common, Southern and Migrant Hawker, but is considerably smaller and, more importantly, has an earlier flight season. Both sexes are black with paired pear-shaped spots down the length of the abdomen – these are mainly blue in the male (he also has blue eyes), and smaller and yellow in the female. The two markings on top of the thorax are narrow stripes in the male, small spots in the female. The thorax in this species is particularly fuzzy but this is not easy to see at any distance. The Hairy Dragonfly is most common in the wetlands of south-east and eastern England, but has patchy distribution elsewhere in Britain, and is widespread in Ireland.

The Hairy Dragonfly is like a miniature version of the larger, summer-flying hawkers – but its flight season does not normally overlap with any of theirs. Otherwise it would surely often fall prey to its larger cousins.

How to find

■ **Timing** This dragonfly emerges in April, making it the first of the 'hawker-like' dragonflies to appear, and flies into early July.
■ **Habitat** The Hairy Dragonfly is most often encountered around still or slow-flowing water with lush reedy margins, and can be common in marshlands. Immature individuals may be found hunting in woodlands.
■ **Search tips** In suitable habitat, scan along ditches and around lake margins for male Hairy Dragonflies as they patrol their territories. From paths that overlook canals or channels, scan the vegetation for resting or newly emerged individuals.

WATCHING TIPS

The male Hairy Dragonfly is lively and aggressive, unafraid to give chase to other species, even the mighty Emperor, and takes large prey, even occasionally dragonflies close to its own size, such as Four-spotted Chasers. Females are relatively shy and observed less often than males but you could spot one laying eggs in a well-hidden corner of the lake or ditch, half-hidden among vegetation. Also check areas like this in the morning for newly emerged Hairy Dragonflies resting before their maiden flight.

Golden-ringed Dragonfly

Cordulegaster boltonii

Length 6.2cm

1	2	3	4	5

J F M A M J J A S O N D

This magnificent dragonfly is unmistakeable given good views, with its green eyes, yellow face, and long black body marked with yellow bands. The male's abdomen is distinctly club-shaped, while the female's is parallel-edged – her elongated, blade-like ovipositer makes her the longest-bodied of all British dragonflies. Golden-ringed Dragonflies are widespread though not usually common. They are absent from Ireland, and in Britain their distribution is concentrated in the North and West, with relatively few colonies in southern and central parts of England.

Look for this beautiful insect in the uplands, on heaths and moors, and woodland edges. Unlike most large dragonflies, it rests often and you can usually get close for a good look. Check the body shape to tell males from females.

How to find

- **Timing** Golden-ringed Dragonflies emerge in June or the very end of May, and fly well into August.
- **Habitat** In its larval stage this species requires acidic water, and populations tend to be based around smaller, partly shaded streams or runnels in upland woods, heaths and moorland. Newly emerged adults spend a week or more away from water, returning to it only when fully mature and ready to mate.
- **Search tips** This dragonfly is not as relentlessly active as the hawkers and you may find it basking on sunlit bracken or a warm rock. Males spend short spells through the day seeking females by flying low over water, but you may also find them hunting well away from water. Females lay their eggs in flight, jabbing their 'tail-tips' into the water, and are very noticeable as they do this.

WATCHING TIPS

This dragonfly rests often, usually quite close to the ground, and can be approached easily when it is perched. Watching a female laying her eggs is impressive – she hovers low over shallow water with her body held almost vertically, and stabs her rear end rapidly and repeatedly into the water, pushing her strong ovipositor into the mud at the bottom of the stream. In the coldest, most northerly parts of its range, the nymphs can take four or more years to mature – look out for them in shallow moorland pools.

Gomphus vulgatissimus

Club-tailed Dragonfly

Length 3.6cm

1	2	**3**	4	5

J	F	M	A	M	J	J	A	S	O	N	D

This is a small and very distinctive dragonfly, named for the bulbous shape of the tip of its abdomen. It is mainly black, with green markings in the mature male, but yellow in females and immature males. It is the only British dragonfly in which the eyes do not touch, giving a wider head shape. The Club-tailed Dragonfly is a rather rare species which depends on large, clean, lowland rivers. This includes several that originate in the Welsh uplands, including the Severn, Dee and Wye, plus the upper reaches of the Thames and some of its tributaries, and the Arun in Sussex. It is a difficult species to see, requiring good luck and correct timing.

It has a unique colour combination and body shape.

Super sites

★ The Thames near Goring, Oxfordshire
★ The Arun near Pulborough, West Sussex
★ The Severn near Bewdley, Worcestershire
★ Slow stretches of the Dee, Severn and Wye, Wales

How to find

■ **Timing** Club-tailed Dragonflies emerge in the first half of May, and their short flight season is over by late June.
■ **Habitat** It is found around lushly vegetated and reasonably undisturbed slow-flowing lowland rivers, and hunts in the surrounding countryside, including fields and woodland edges.
■ **Search tips** The best way to see the species is to visit at the start of the season and look for freshly emerged individuals. Because most larvae will leave the water within the same few days, at large colonies you should find at least a few. Search carefully at the river's edge, in surrounding grassland and on shrubs. Look for the moulted larval skins (exuviae) to confirm that emergence is underway. Finding the mature insects is more difficult but watch for males flying low over the water or perching on overhanging vegetation.

WATCHING TIPS

If you are lucky enough to find a freshly emerged Club-tailed Dragonfly, you should be able to examine it at length as it will be reluctant to fly for at least an hour or two, and will fly only weakly for some time after that. Watching territorial and breeding behaviour is difficult, much more so than finding newly emerged immature insects, but you could spot males flying in low, wide circles over the water, or females egg-laying in flight with rapid dipping motions of their abdomens.

Downy Emerald

Cordulia aenea

Length 3.7cm

1	2	3	4	5

J	F	M	A	M	J	J	A	S	O	N	D

This is the most common and widespread of the three emerald dragonfly species that occur in Britain, though it is still rather localised. Its stronghold is southern and central England, including parts of Greater London, but it also has small populations in East Anglia, south-west and north-west England, north-west Scotland, and south-western Ireland. It is medium-sized with an unmarked shiny green body and bright green eyes. The male's abdomen is club-shaped when seen from above, and in profile looks rather arched and flattened. It has a bronzy sheen, especially towards the tail-tip. The female has a greener and straight-sided body.

Super sites

- ★ The Moat Pond, Thursley Common, Surrey
- ★ Sevenoaks Wildlife Reserve, Kent
- ★ Wimbledon Common, Greater London
- ★ Wraymires Tarn, Cumbria
- ★ Coire Loch, Scottish Highlands

How to find

■ **Timing** Downy Emeralds fly between mid-May and early July. They are easiest to see in the middle of the day when males are patrolling their territories – at other times they are often resting high in trees.

■ **Habitat** These dragonflies like sheltered pools and lakes within woodland, with some sunny and shady areas, and usually slightly acidic water.

■ **Search tips** The males are easy to spot when they are patrolling their territories, working their way along a stretch of lake shore and often chasing each other about. Warm days with sunshine and little breeze are best. When the sun goes in, the dragonflies will soon fly off to rest away from the waterside.

WATCHING TIPS

Male Downy Emeralds give good views (binoculars are helpful) as they fly along the lake shore in a distinctive down-tilted posture, often moving into inlets and hovering for a second or two before moving on. You will also see them having high-speed clashes with other males, but if they locate a female the pair quickly fly up into the treetops to mate; the same applies if they capture a large prey item. Often a new male will quickly move into the vacated territory. Females are rarely seen but look out for them egg-laying in flight in shallow, sunny waters.

Somatochlora metallica

Brilliant Emerald

Length 4cm

| 1 | 2 | 3 | **4** | 5 |

J F M A M J J A S O N D

Both sexes are an intense, iridescent green.

Super sites

* ★ The Moat Pond, Thursley Common, Surrey
* ★ Old Lodge Nature Reserve, West Sussex
* ★ Coire Loch and other lochs in Glen Affric, Scottish Highlands

The British distribution of this dragonfly is very curious – it occurs in a small area of south-east England, and another small area in north-west Scotland, but nowhere in between. In England it occurs at some of the same sites as Downy Emerald, and in Scotland its distribution overlaps with Downy and Northern Emeralds, so identification can be an issue. Try to see the front of the face, which in this species has a broad yellow band between the eyes – Downy has no yellow here, and Northern just small yellow spots on the side that don't meet in the middle. Seen closely, the female has a distinctive feature – a long vulval scale near the tail tip which points down at right angles to the body.

How to find

■ **Timing** The Brilliant Emerald flies later than the Downy, but there is overlap; it emerges in mid-June and flies into August.

■ **Habitat** It occurs mainly around fairly large, slightly acidic lakes, usually in woodland, and also on some slow-flowing shady rivers. It may be seen hunting in surrounding woodland or heathland.

■ **Search tips** Male Brilliant Emeralds can be found as they patrol lake shores on sunny days. In slightly cooler conditions, look for females discreetly egg-laying near shaded parts of the shoreline – you may also find females resting or hunting in glades or sunny open ground near the breeding lakes.

WATCHING TIPS

As with Downy Emeralds, male Brilliant Emeralds patrol a lakeside territory for a short spell, before departing into the trees to rest or hunt (or, if they have managed to find a female, to mate). They rarely if ever hunt when over the water. At sites with both species, compare their appearance in flight – the Brilliant is brighter, but also has a different, more horizontal position in flight. At rest, this dragonfly is simply stunning and it is worth exploring the areas around the lakes to try to find females hunting or resting.

Northern Emerald

Somatochlora arctica

Length 3.7cm

| 1 | 2 | 3 | **4** | 5 |

J F M A M J J A S O N D

Northern Emeralds occur in northern and mountainous areas across western Europe.

This species has a very northerly distribution in Britain and Europe. It is found mainly in the north-west highlands of Scotland, though there is also a small population in south-west Ireland. It is the darkest and dullest of the emeralds, its vivid green eyes being the brightest part of its body, but has the typical emerald body shape, with the male's abdomen markedly club-shaped, the female's parallel-edged. It is rather a difficult species to observe, and weather conditions in its habitat are highly unpredictable, adding to the challenge.

Super sites

★ Coire Loch and other lochs in Glen Affric, Scottish Highlands
★ Bridge of Grudie near Loch Maree, Scottish Highlands

How to find

■ **Timing** Northern Emeralds fly from early June to mid or late August.
■ **Habitat** This species is associated with small pools and lakes in open moorland and patchy, open woodland. Its preferred pools have floating sphagnum moss, which provides shelter for the growing larvae.
■ **Search tips** As with the other emeralds, it is easiest to see the males when they are flying around the lake shores, hoping to encounter females – though you will need a sunny interval for this to happen, and weather in its habitat is highly changeable. A careful search around the margins may reveal a pair joined 'in cop', although they usually retreat to nearby treetops to mate.

WATCHING TIPS

At some Scottish sites you could find all three emeralds 'working' the same loch. Allow some time to sort out identification, and get familiar with the slight differences in their behaviour and appearance in flight. When the sun shines, activity around the lakes can be frantic and confusing, with Common Hawkers and Golden-ringed Dragonflies also joining the fray, and the spiralling chases between rival males are impressive to watch.

Sympetrum striolatum, S. sanguineum   **Common and Ruddy Darter**

Length 2.6cm

| 1 | 2 | 3 | 4 | 5 |

J F M A M J J A S O N D

Common Darter (male)

Males of these two species are the small red dragonflies that are common in many places through mid-summer and autumn. The Ruddy is brighter red, and has a more club-shaped abdomen than the Common. Females are yellow-brown and harder to tell apart, but a reliable way is to check the legs. Commons have an obvious yellow stripe, while Ruddys are completely black. Common Darters in north Scotland have more black coloration and are sometimes treated as a separate species – the Highland Darter. The Ruddy Darter is found across most of England but becomes scarce in the South-West and North. It also occurs in south Wales and widely in Ireland. The Common is widespread throughout the British Isles.

Ruddy Darter (male)

On cooler days it is not unusual for darter dragonflies to settle on passing humans, benefiting from a little extra warmth. As autumn draws to a close the last few Common Darter pairs may be seen frantically egg-laying in small ponds.

How to find

■ **Timing** Both of these darters emerge in July, and fly into autumn, the Ruddy tending to disappear first, while small numbers of Common Darters survive into November if the weather is mild enough.

■ **Habitat** Both of these darters breed on all kinds of still and slow-flowing waters, from garden ponds and ditches to large lakes and rivers, and immature insects may be found in woodlands, farmland and open countryside some distance from water.

■ **Search tips** Darters often settle on the ground – you may disturb them as you walk along, but they will soon land again. On cooler days they will warm up on wooden fences and will even land on people's hands. When the air temperature is higher, they may choose higher, tree-top perches.

WATCHING TIPS

These darters are very approachable – easy to find and easy to watch. They watch out for prey and potential mates from their perch, flying out when something catches their attention, but usually landing again nearby, and as long as you approach slowly you should be able to get close enough for a really good look. Pairs stay joined in the 'tandem' position (the male using his claspers to grip the female behind her head) in flight while the female lays her eggs, and in good conditions you may see dozens of joined pairs flying low over the surface, repeatedly dipping down to the water.

White-faced Darter

Leucorrhinia dubia

| 1 | 2 | 3 | 4 | 5 |

J F M A M J J A S O N D

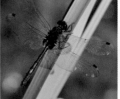

This is a delicate-looking dragonfly with unique colours and markings. The male is primarily black, with some square red markings on the abdomen and red areas on the thorax. The female is black with yellow markings, as is the immature male. There is a small black patch at the base of the hindwings, and the eyes are black and face bright white in both sexes. This is a rare and very localised species, with exacting habitat needs. It is mainly found in north-west Scotland, with other populations in upland parts of north-west and west-central England – the small outpost at Thursley Common in Surrey has, sadly, disappeared.

No other British dragonfly shows such vivid red and black coloration.

Super sites

★ Whixall Moss, Shropshire
★ RSPB Abernethy Forest, Highland (Tulloch Moor)
★ Glen Affric, Scottish Highlands

How to find

■ **Timing** Appearing earlier than most other darters, this species flies from May into late July or early August. It is active on warm sunny days.
■ **Habitat** It occurs on moorland with areas of trees or scrub, in the vicinity of acidic, boggy pools with sphagnum moss.
■ **Search tips** Males may be found as they patrol their waterside territories; also look for both males and females basking on logs or rocks near or away from the water. Look for newly emerged individuals hunting in nearby patches of woodland.

WATCHING TIPS

You can watch males flying low over the water in the sunshine and pausing to rest, as they wait for females to arrive. Once a pair meets and the male takes hold of the female with his claspers, they usually settle on the ground near the water to copulate. Unlike with Common and Ruddy Darters, the pair separate before the female lays her eggs among floating sphagnum moss, though he may stay nearby, watchful for other intruding males.

Sympetrum danae

Black Darter

Length 2.4cm

1	2	3	4	5

J F M A M J J A S O N D

One of the smallest dragonflies, the Black Darter is a distinctive and attractive species. The male is almost completely black when seen from above, with some yellow markings on his sides, while the female is yellow with much black on her sides and underside. Immature males resemble females but can be identified by their more club-shaped abdomens. The Black Darter occurs widely in Scotland, north-west England, Wales, south-west England and much of Ireland. It is mostly absent from south-east England though has strong populations on Surrey and Hampshire heathlands.

This small and delicately built dragonfly is often targetted by bigger and fiercer 'dragons', and it usually avoids flying over open water in the middle of the day when Emperor Dragonflies are on the prowl.

How to find

■ **Timing** This species emerges in early July and flies into mid-September.

■ **Habitat** It occurs around well-vegetated acidic ponds and lakes on open ground, primarily on boggy heaths and moors in the uplands and lowlands.

■ **Search tips** Look out for these dragonflies basking on the ground or on vegetation around the breeding ponds in the mornings. Later in the afternoon there is more activity over the water, with males chasing unpaired females.

WATCHING TIPS

Black Darters are not flustered by close scrutiny so are easy to watch and often choose conspicuous, exposed perches. Their activity peaks in the warm afternoon hours, when you may see multiple males trying to catch a female, and also pairs in tandem flying over the water, dipping down as the female lays her eggs. Sometimes females will lay eggs by themselves, with or without the male keeping watch for other males, and for potential predators. Being so small, this species quite often falls prey to other dragonflies as well as more expected dangers.

Red-veined Darter

Sympetrum fonscolombii

Length 2.8cm

1	2	3	4	5

J F M A M J J A S O N D

This rather enigmatic species is mainly a migrant visitor to Britain but does regularly establish colonies, some of which persist for years at a time. It is the only red darter species likely to be seen in Britain in springtime. Males are very red, with red wing-veins, while females are yellow, with yellow wing-veins. The body is straight-sided and a little stouter than in the Common Darter in both sexes. One of the best identification traits is revealed when you look at it from the side – in both sexes the lower part of the eye is blue. This darter is most likely to be encountered in England, along southern and eastern coasts.

If you regularly visit southern Europe, you may well know this species – it is very common in the Mediterranean region. Climate change may assist its colonisation of Britain.

How to find

- **Timing** Red-veined Darters are often reported as early as May, and through into summer and early autumn, though later individuals may be overlooked among the many Common and Ruddy Darters on the wing at that time.
- **Habitat** It breeds in all kinds of shallow, usually well-vegetated waters.
- **Search tips** Males are territorial and perch alertly near the water, ready to chase off other males or attempt to grab females. In the cooler hours they will perch on rocks or bare ground for warmth, but move to vegetation when temperatures rise.

WATCHING TIPS

This unpredictable dragonfly is potentially findable by any wildlife-watcher over most of Britain – in 2017, several adults were present in Kensington Gardens in central London. It is worth checking all red darters just in case, but especially those found in spring and early summer. These are very likely to be Red-veined. It is an approachable insect and obtaining good views and photos is often possible. Any sightings should be passed on to your local recorder.

Libellula depressa

Broad-bodied Chaser

Length 3cm

| 1 | 2 | 3 | 4 | 5 |

J F M A M J J A S O N D

The Broad-bodied Chaser often chooses the very tip of a bare, almost vertical stem as its vantage point, from which to scan for prey and possible mating partners. After repeated matings the male's blue dusting wears away, leaving dark 'scars'.

This is a striking dragonfly, with its robust body and habit of constantly returning to the same often exposed and elevated perch. Freshly emerged individuals of both sexes are quite bright orange-brown. With maturity, males' abdomens become powder-blue as a dusty 'bloom' (pruinescence) gradually develops – older males develop worn black patches on the abdomen sides. Females (which are noticeably wider-bodied) become a little darker and greener with age. Both sexes have yellow markings on the abdomen sides, and black patches at the base of each wing. The face and eyes are dark. Broad-bodied Chasers are widespread in most of England and Wales, but rare in north England and absent from Scotland and Ireland.

How to find

■ **Timing** Adults appear in May, and are among the first dragonflies on the wing each year. They fly into late July or early August.

■ **Habitat** Broad-bodied Chasers can breed at all kinds of still and reasonably clean water, from smaller garden ponds to large lakes – they prefer fairly open water rather than that with extensive vegetation at the margins. They will hunt in nearby woodlands and fields.

■ **Search tips** Look out for this dragonfly in wooded areas on sunny days from mid-May, and around water a couple of weeks later. It is very eye-catching, both in flight and at rest.

WATCHING TIPS

This species and other chasers have the very convenient habit of returning again and again to the same perch, and resting for long spells in between short chases in pursuit of prey or (in the case of males) a passing female. Resting individuals can usually be approached closely. Mating happens in mid-air, a rapid, wing-clattering tussle, after which the female lays eggs by dipping her body in the water in flight, while the male stands guard nearby.

Four-spotted Chaser

Libellula quadrimaculata

Length 3cm

1	2	3	4	5

J F M A M J J A S O N D

In a few individuals, the dark pterostigma markings at the wingtips are extended into dusky bands. This attractive colour variant is known as *praenubila*, and is fairly frequent in some populations.

This is a common dragonfly, reaching very high numbers at particularly good sites. It is also very widespread, occurring across almost all of Britain and Ireland. It is named for the dark spot halfway along the leading edge of each wing, and there is also a dark patch at the base of each wing, as well as the coloured cell near the wingtip (the pterostigma), present in almost all dragonflies and damselflies. Otherwise, it is rather plain, with its dark, tapering grey-green body marked only with pale yellow patches on the sides. The eyes are dark, the face pale. The shape of the claspers (flared in the male, coming together in the female) is the most obvious difference between the sexes.

How to find

- **Timing** Four-spotted Chasers emerge in late May, and fly until mid-August.
- **Habitat** This adaptable species can be found in all kinds of habitats with some fresh still or slow-flowing water, from heaths and moors to marshland and woodland pools, open gravel pits and reedy ditches.
- **Search tips** Four-spotted Chasers are quite active but will return to favourite perches in between territorial chases or hunting forays. They often vastly outnumber other dragonfly species and at high densities their activity can be frenetic – where numbers are lower, look out for them perched on high vegetation overlooking the water.

WATCHING TIPS

Where many male Four-spotted Chasers are battling for territorial rights over one small area of open water, you'll witness almost constant chases and battles – elsewhere in the species' vast global range, overly high numbers can trigger mass migrations. Such events are rare in Britain, though wanderers do turn up far from suitable habitats. Females lay their eggs in flight, guarded by their most recent mate, who may attempt to mate again immediately after the female has finished laying.

Libellula fulva

Scarce Chaser

Length 3cm

1	2	3	4	5

J F M A M J J A S O N D

Although much more localised than our other two chasers, this river-dwelling species is expanding its range from its strongholds in south and eastern England. When newly emerged, it is strikingly orange with a bold black line down its abdomen, and a smudgy dark patch at each wingtip. With age, the female becomes browner, while the male develops a light blue abdomen with a black tip, and light blue eyes; his wingtip patch often fades completely away. There are small dark patches at the base of each wing in both sexes. The Scarce Chaser somewhat resembles the Black-tailed Skimmer – look for the basal wing patches, the male's blue eyes, and the female's black abdomen stripe and dark wingtips.

This charming small dragonfly seems to be becoming less scarce each year.

Super sites

★ Stour Valley, Kent
★ River Arun, West Sussex
★ RSPB Strumpshaw Fen, Norfolk

How to find

■ **Timing** Scarce Chasers are on the wing from mid-May until late July.

■ **Habitat** This species occurs around slow-flowing rivers and adjacent channels, ditches and water meadows, with plenty of vegetation around including some trees and shrubs.

■ **Search tips** Like other chasers, it selects favourite perches and returns to them repeatedly, so if you disturb one from its perch, wait and it may return. Males perch alongside a stretch of riverside from mid-morning, often quite high up – scan overhanging branches with binoculars.

WATCHING TIPS

Males monitoring their territories make frequent sorties to investigate other passing insects in the hope they may be prey or a potential mate. Females tend to hunt away from water until they are looking to mate, and in between hunting flights they bask on a high perch. Pairs couple in the air but often land near the water and stay attached for several minutes. The female then lays her eggs with typical dipping actions while the male waits nearby.

Keeled Skimmer

Orthetrum coerulescens

Length 2.9cm

1	2	3	4	5

J F M A M J J A S O N D

This is a small and very slim-bodied dragonfly, which could even be mistaken for a blue damselfly at first glance. The mature male's abdomen is entirely powder blue, while female and immature males are straw-coloured with narrow black markings. They resemble female darters but are more robust and have a dark thorax top with two prominent pale stripes. The wings are unmarked apart from the yellow pterostigmata near the wingtips. This species has a very westerly bias in its British distribution, occurring in Wales, south-west and north-west England, the far west of Scotland, and on both coasts in Ireland; there are also isolated outposts in parts of central and eastern England.

The wings are often swept well forwards when the dragonfly is at rest (main image).

Super sites

★ New Forest, Hampshire
★ Thursley Common, Surrey
★ Cors Caron NNR, Cardiganshire

How to find

■ **Timing** This species emerges in the first half of June, and is on the wing until late August.
■ **Habitat** Keeled Skimmers prefers open boggy or heath habitats with small pools and channels that have floating sphagnum moss.
■ **Search tips** This species is often noticed as it basks on warm ground or rocks close to the water. Males holding territory will also patrol on the wing as well as watching from a perch.

WATCHING TIPS

The Keeled Skimmer spends long spells resting on sun-warmed ground with its wings turned forward to expose the body to heat as much as possible. Males hold small territories and seize a female when she arrives. Although females lay their eggs without the male attached, before mating the pair may fly in tandem over the water, the male perhaps showing the female potential egg-laying sites in his territory. As with all Odonata, a male can take hold of a female without her co-operation, but she chooses whether or not to complete the act by curving her abdomen under her own body to meet his and form the mating 'wheel'.

Orthetrum cancellatum

Black-tailed Skimmer

Length 3.5cm

1	2	3	4	5

J F M A M J J A S O N D

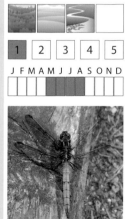

This skimmer is common in the southern half of England, and occurs in south Wales, central Ireland, and more sparsely in central and northern England. The mature male has a blue abdomen with a black tip and orange-yellow patches along its sides. The wings are entirely clear except for the pterostigmata, which helps distinguish it from Broad-bodied and Scarce Chasers; it is also slimmer than any chasers. The eyes are dark, the face pale. Females and immature males are yellowish with two black lines down the abdomen, either side of the centre. Very old females darken to almost black.

This dragonfly changes in appearance quite dramatically through its adult lifetime, with both sexes showing obvious signs of age after a few weeks of life, but its clear wings, shape, size and behaviour always help distinguish it from other species.

How to find

■ **Timing** This species emerges at the end of May, and flies into early August.

■ **Habitat** Various types of water body support this skimmer, from small ponds to rivers and marshes. It prefers waters with some bare ground at the shoreline.

■ **Search tips** This skimmer most often rests on the ground, rarely settling on a raised perch of any kind. It is quite flighty – you may first notice it when it takes off in front of you, or flies low over the water. If you spot a male hovering over water, look out for a female laying her eggs nearby, hovering low and stabbing her abdomen rapidly into the water.

WATCHING TIPS

This skimmer is a powerful predator – watch one for a while and you may see it tackle a damselfly or butterfly. Males are also fiercely defensive of their territories. Females stay away from water until ready to mate, and as they visit possible egg-laying sites they will quickly be approached and seized by a male. After mating, a female lays her eggs alone, but sometimes will be grabbed by another male, who will attempt to empty her genital tract of existing sperm so that she will mate again.

Banded Demoiselle

Calopteryx splendens

Length 3.8cm

| 1 | 2 | 3 | 4 | 5 |

J F M A M J J A S O N D

This damselfly is one of the joys of a riverside walk on a warm sunny day. It is longer in the body than several smaller dragonfly species, but has the unmistakable damselfly shape, with wide-spaced eyes forming a dumbbell-shaped head, and a long, slim, parallel-edged body. The male is deep metallic blue, with a large dark patch on the outer half of each wing. The female is metallic green; her wings are unmarked but have a strong green tint, and a contrasting white spot at each wingtip (this is lacking in immature males, which otherwise resemble females). Banded Demoiselles are very common in south and central England, Wales and Ireland, and sparingly present in north-east England.

The vivid colours of the Banded Demoiselle, and the male's rather extrovert habits, make it easy to spot and to watch. Take a picnic to your local riverside on a warm sunny lunchtime, and enjoy the show.

How to find

■ **Timing** The first individuals appear in late May. More emerge through summer and the flight season extends into September.

■ **Habitat** This is a species of slow-flowing, lushly vegetated rivers. It may also be found in fields and glades nearby.

■ **Search tips** Male Banded Demoiselles draw attention with their showy, fluttering flight on large, strongly marked wings, and the wing-flicking display they perform while perched – often there will be several 'performing' in close proximity. Activity is highest in the warmer hours of the day. In the evening, they often go to roost in small groups, in long vegetation near the water.

WATCHING TIPS

Watching the males dancing on the riverside is delightful. They display around suitable egg-laying sites, and when a female arrives they draw her attention to the spot with downward swoops and perching with the wings flicking open and shut – if she is impressed she allows the male to mate. When there are many males and fewer good sites, things become more competitive and males try to grab females in flight, or dislodge males that are already paired. These are quite skittish damselflies, easily disturbed – but later in the day individuals of both sexes go to roost away from the water's edge when they are more approachable.

Calopteryx virgo

Beautiful Demoiselle

Length 3.8cm

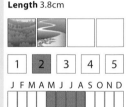

1	2	3	4	5

J F M A M J J A S O N D

This stunning insect is similar to the Banded Demoiselle, but the wings are broader and entirely dark purple-blue in the mature male; the female's are bronze-green, and the outer part of her abdomen is also bronze. In flight, the male looks like a dark butterfly; when seen well, the abdomen is glossed deep blue-green. This damselfly has slightly different habitat needs than the Banded Demoiselle so although their distributions overlap, many sites support one species but not the other. This damselfly's British distribution is concentrated in the south-west of England, Wales and south Ireland, more sparsely in central and south-east England, with some colonies in north and west England and west Scotland.

The female's wings are a lovely golden-bronze colour, and broader than those of the female Banded Demoiselle. Immature males look rather female-like, but only females show the white wingtip spots ('pseudopterostigmata').

How to find

■ **Timing** This species is on the wing from mid or late May until late August.

■ **Habitat** Beautiful Demoiselles are associated with clean, fast-flowing streams and rivers with gravelly or sandy bottoms, often in or near woodland.

■ **Search tips** As with the Banded Demoiselle, males of this species perform an eye-catching display with a dancing flight and perched wing-flicks, close to the water's edge. Both sexes also often perch quite prominently on plant foliage a metre or two above ground level.

WATCHING TIPS

These damselflies tend to be less flighty than Banded Demoiselles, so can be approached and studied closely. The courtship behaviour is fascinating. When a female approaches a male at the water's edge, he performs a fluttering dance around her, and flies down to the water to show her where she might lay eggs. The pair then copulate, during which the male's secondary genitalia is used to empty out any sperm already in the female's genital tract before adding his own. After mating, the male stays near the female while she lays her eggs into submerged water plants.

Emerald Damselfly

Lestes sponsa

Length 3cm

1	2	3	4	5

J F M A M J J A S O N D

This is the most common British representative of the *Lestes* damselflies, known as emeralds in Britain but 'spreadwings' elsewhere. This name describes their typical resting posture, with wings held open at about 45 degrees to the body, rather than closed over the abdomen as in other damselflies. It has a metallic green body, which in the male has patches of pale blue pruinescence at the base and tip of the abdomen. He also has blue eyes, while the female's are brown. This species is found very widely across Britain and Ireland, although is often not particularly common, and is missing from some sites that appear suitable.

The emerald damselflies are relatively large. Females are also quite robust with parallel-edged bodies, but males are needle slim in their mid-section, with a more bulging abdomen tip.

How to find

■ **Timing** The first individuals appear in late June, and emergence continues through summer, although individuals may also live for several weeks. The flight season ends in late September.

■ **Habitat** It is found around all kinds of still water, from small ponds and ditches to large lakes, but has a particular affinity with boggy acidic pools on open heathland.

■ **Search tips** Look out for Emerald Damselflies perched on sedge stems or other waterside vegetation, the sunlight making their bodies gleam brightly.

WATCHING TIPS

This beautiful, iridescent damselfly is usually quite approachable. You will see pairs resting or flying in tandem in the warmer part of the day, and egg-laying occurs with the male still attached, a potential lifeline for the female as she often submerges fully for minutes at a time. Conversely, she may lay eggs above the water-line of partly dried-out pools – the eggs do not hatch until the following spring, and can survive out of water for months.

Scarce and Willow Emerald Damselflies

Length Scarce Emerald 3cm
Willow Emerald 3.5cm

1	2	**3**	4	5

J F M A M J J A S O N D

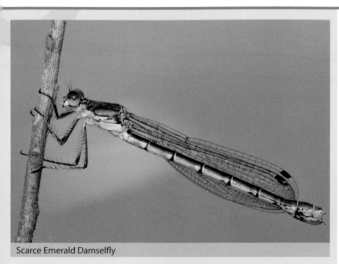

Scarce Emerald Damselfly

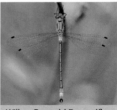

Willow Emerald Damselfly

Several other *Lestes* species have been recorded in Britain besides the Emerald Damselfly. The Scarce Emerald is found only in south-east England and parts of Ireland, while the Willow Emerald is a new arrival, arriving in parts of east and south-east England in the early 2000s and quickly spreading north and west. The Scarce Emerald closely resembles the Emerald – males have slightly less blue at the top of the abdomen, and the female has two square spots at the very top of the abdomen (these are rounded in the Emerald). The Willow Emerald is larger, and males have no blue at all. The pterostigma is light brown rather than black, and there is a spur-shaped marking on the thorax side.

The Willow Emerald is now probably more common than the Scarce Emerald, and any English wetland site with willows and ideally some slow-flowing water is worth checking in late summer for this species.

How to find

■ **Timing** The Scarce Emerald Damselfly flies from mid-June to mid-August. The Willow Emerald appears from mid-July and may linger as long as October, making it one of our latest-flying Odonata.

■ **Habitat** The Scarce Emerald favours shallow small pools and ditches, fresh or brackish, that are inclined to dry out in summer. The Willow Emerald uses ponds and slow-flowing rivers with willows growing along the banks.

■ **Search tips** Look out for these damselflies perched near suitable breeding waters, with wings half open in the typical style. Scarce Emeralds take some searching out as they often rest within dense vegetation. Willow Emerald females lay their eggs into willow branches and the laying sites are visible as neat rows of small scars. Look out for this sign at areas near where Willow Emeralds are becoming established, to help chart the spread of this new British species.

WATCHING TIPS

Scarce Emeralds may be seen flitting around the shores of their small breeding pools – look out for tandem pairs settling above the water line for the female to lay her eggs. Female Willow Emeralds lay eggs without the male attached, landing on vertical branches and bending their abdomens almost double to lay an egg, then shifting position to lay the next. Both of these species could turn up outside their currently known range – let your local recorder know of any sightings in new areas.

White-legged Damselfly

Platycnemis pennipes

Length 3.1cm

1	2	3	4	5

J F M A M J J A S O N D

A side-on view reveals the thickened, feather-like leg structure, and also the double-stripe on the abdomen sides that distinguishes this rather unusual species from the other blue damselflies.

This is an unusual damselfly, the only British representative of the family known as 'featherlegs'. It is rather a pale damselfly – light blue in mature males, creamy in females and immature males, with narrow black markings – these are very sparse in newly emerged individuals of both sexes. The legs are white and thickened, especially the hind pair, and the head is exceptionally wide across. The thorax side bears a narrow double black stripe. This damselfly is fairly common and widespread in southern and central England, with strong populations in the New Forest and Thames Valley.

How to find

■ **Timing** This species flies from late May into early August.
■ **Habitat** White-legged Damselflies are mostly found around slow-flowing rivers and streams with plenty of emergent and floating vegetation. They have disappeared from some stretches of river where the backwash from frequent boat traffic has reduced waterside vegetation.
■ **Search tips** This is a sluggish damselfly, which can be found resting or flitting about close to the water's edge, later in the day resting on sedge stems or other vertical plants.

WATCHING TIPS

Getting good views of White-legged Damselflies is fairly easy as they are not as easily startled as most other species. If you lean in towards one on a stem, it will often shimmy around to the other side – its very wide-set eyes still studying you comically on either side of the stem. Pairs in tandem and unpaired females are drawn to emerging water plants, especially water lily flowers, and perch on these to lay eggs, shimmying down the stem to dip their abdomen tips into the water.

Enallagma cyathigerum

Common Blue Damselfly

Length 2.8cm

| 1 | 2 | 3 | 4 | 5 |

J F M A M J J A S O N D

This is the most widespread of our bright blue damselflies, and although it looks similar to the other five species it is in fact not closely related to them. The male is blue with black markings, the female dull brown or sometimes blue with much more extensive black markings. The easiest way to tell both sexes from other damselflies is the thorax-side pattern, with a broad pale stripe above a narrow black one, and without the short black 'spur' below that is found on all four *Coenagrion* species. Common Blue Damselflies occur throughout Britain and Ireland. Those in north Scotland have more extensive black markings than those elsewhere.

This very abundant damselfly has a longer flight season than most other small 'damsels', often hanging on into autumn. Despite its similarity to other blue damselflies, it is actually the sole British representative of an otherwise American genus.

How to find

■ **Timing** It is on the wing from April through to October.

■ **Habitat** This species can be found in almost any kind of habitat with fresh water, from a small garden pond in London to a moorland loch in the Highlands of Scotland. Newly emerged adults often hunt in fields and woodland edges some distance from water.

■ **Search tips** As you walk in suitable habitat you'll notice these damselflies (and probably other species) lifting up from the vegetation, and in the middle of the day there will probably be pairs in tandem, on vegetation and flying over the water. They resemble moving, coloured needles, as the clear wings are hard to see in flight, except in the case of newly emerged individuals, which have eye-catchingly shiny wings.

WATCHING TIPS

You can see all elements of the adults' lives in a single day by the waterside. Like most small damselflies, this species perches and waits for prey, before flying out to catch it. Mating and egg-laying usually takes place in the middle of the day, and later, as the sun sets, the insects look for safe, sheltered roosting places in long grass or other vegetation. They are often caught in spiders' webs, and if no spider is in attendance then rescuing a trapped damselfly gives a great opportunity for you to examine its delicate beauty close-up.

Azure and Variable Damselfly

Coenagrion puella, C. pulchellum

Length 2.8cm

| 1 | 2 | 3 | 4 | 5 |

J F M A M J J A S O N D

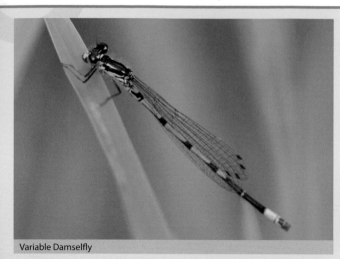

Variable Damselfly

Azure Damselflies mating

These two species are closely related and very similar. Males of both are blue, while females have a green form and a blue form. The male Azure has less black on its abdomen and thorax, and the marking on the abdomen just behind the thorax is shaped like a narrow U, while the male Variable's marking here is a thicker U with a stalk. Both can be distinguished from the Common Blue by this (the Common Blue's marking here is shaped like a small mushroom), as well as by the black spur marking on the thorax side. Females of the two species, though, are very difficult to separate. The pronotum, which joins the head to the thorax, is differently shaped in the species – good close-up photos show this.

The Azure Damselfly is common throughout Britain except north Scotland. The Variable is most common in the south and east but is common and widespread in Ireland. At several sites both will occur, though one tends to predominate.

How to find

- **Timing** Both species fly from late May to early August.
- **Habitat** They are found around all kinds of well-vegetated still water.
- **Search tips** The advice given for finding the Common Blue applies to these species as well. Both have a shorter flight season than the Common Blue and, in June when numbers peak, they will usually greatly outnumber Common Blues.

WATCHING TIPS

At good sites, the number of tandem pairs you'll see of these two species can be extraordinary, all crowding around the best egg-laying sites, with unpaired males doing their best to intervene with the joined-up couples. Identification of Azure and Variable Damselflies is difficult but the average smartphone can take photos clear enough to show the key features. Immatures of both of these species can be found in profusion in grassland close to water – they do not return to the waterside until fully mature and ready to mate.

Coenagrion mercuriale,
C. hastulatum, C. lunulatum

Southern, Northern and Irish Damselflies

Length 2.6cm

1	2	3	**4**	5

J F M A M J J A S O N D

Irish Damselfly

Southern Damselfly

Our remaining three *Coenagrion* damselflies are all highly localised. The Southern is present patchily in south and south-west England, south Wales and Anglesey, the Northern occurs in the north-central highlands of Scotland, and the Irish is present only in Ireland, mainly in northern and central parts. Males of all three have more black on their abdomens than the Azure, Variable and Common Blue, and all three species are smaller than the commoner blue damselflies. The marking at the top of the abdomen in the male is distinctive in all three – a U balanced on a mushroom in the Southern, a playing-card spade symbol in the Northern, and a horizontal crescent moon shape in the Irish.

Super sites

* Crockford stream, Hampshire (Southern)
* Somerset Levels, Somerset (Southern)
* RSPB Abernethy Forest, Highlands (Northern)
* Dinnett NNR, Deeside (Northern)
* Montiaghs Moss Nature Reserve, Co. Armagh (Irish)

How to find

■ **Timing** These species fly from late May to the end of July or start of August.
■ **Habitat** The Southern prefers slow-flowing streams and rivers on chalky soil, the Northern is found around small lush woodland pools, and the Irish occurs around clear, peaty moorland pools.
■ **Search tips** June or early July is the best time to seek out all of these species. If you have a reliable site and a sunny day, sightings are very likely – check on and around vegetation at the water's edge. They are all easiest to find on sunny, not too breezy days, towards the middle of the day when courtship and mating behaviour peaks.

WATCHING TIPS

You may well need to make a special trip to see one or all of these species, so take the time to make a certain identification. With Northern, identification is relatively easy as the only confusion species is Common Blue, but the other two could occur alongside Azure or Variable Damselflies. These insects are usually too flighty to allow you to inspect them with a hand lens, so try to take some clear photos when you need to be certain of identification.

Blue-tailed and Scarce Blue-tailed Damselfly

Ischnura elegans,
I. pumilio

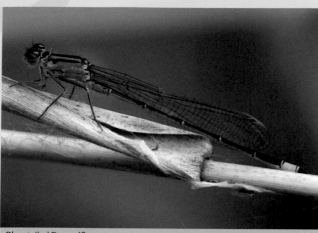

Blue-tailed Damselfly

Length 2.5cm

1	2	3	4	5

J F M A M J J A S O N D

Scarce Blue-tailed Damselfly

The Blue-tailed Damselfly is, along with Common Blue, our most common and widespread damselfly, occurring throughout Britain and Ireland. Its very similar relative, the Scarce Blue-tailed, lives up to its name in being uncommon, localised and elusive – it is present patchily in southern and south-west England, south Wales and parts of Ireland. Adult males are near-identical, differing in the size of the blue marking at the tip of the otherwise black abdomen. Females, though, are more easily separated – in fact, the immature female of Scarce Blue-tailed is one of the most distinctive and attractive of all our damselflies, with its bright golden-red and black pattern.

The blue-tailed damselflies are extremely small and dainty. Female Blue-taileds of the *typica* form resemble males but are slightly thicker-bodied, and have a just-visible ovipositor at the abdomen tip.

How to find

■ **Timing** Both species appear in late May. The Scarce Blue-tailed's season finishes mid-way through August, while the Blue-tailed can still be seen into October.
■ **Habitat** The Blue-tailed occurs in proximity to all kinds of fresh water. The Scarce Blue-tailed prefers shallow, often temporary pools in bare and disturbed ground – even puddles on country paths may attract them.
■ **Search tips** These damselflies are small and rather inconspicuous, and rarely occur at high densities – you may need to search carefully for some time, especially for the Scarce Blue-tailed.

WATCHING TIPS

The Blue-tailed Damselfly has a number of distinct immature and mature female forms, and finding and documenting all of these is an interesting exercise. Immatures have either a violet thorax (*violacea*) or a rosy-pink one (*rufescens*). Form *violacea* may mature into the male-like *typica*, with a black, blue-tipped abdomen, or the greenish *infuscens*. Form *rufescens* becomes the greenish-brown *rufescens-obsoleta*. Scarce Blue-tailed's reliance on transient habitats means colonies shift from year to year – your local recorder will be interested in all observations.

Erythromma najas,
E. viridulum

Red-eyed and Small Red-eyed Damselfly

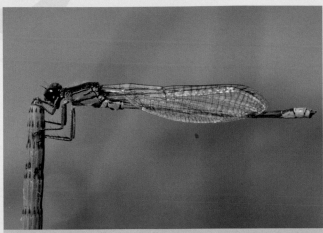
Small Red-eyed Damselfly

Length Red-eyed 3cm
Small Red-eyed 2.5cm

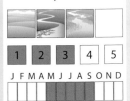

1	2	3	4	5

J	F	M	A	M	J	J	A	S	O	N	D

Red-eyed Damselfly

Males of both these species have bright red eyes that are striking even at some distance. The male is otherwise black with a blue tail-tip – this bears a narrow cross-shaped marking in the Small Red-eyed. Females are dark and dull, with dark but very beady-looking eyes. The Small Red-eyed female has a pair of pale stripes on the thorax top – in the Red-eyed these are reduced to spots. The Red-eyed Damselfly is common over most of England except the North and far South-West, and also in east Wales. The Small Red-eyed is a recent colonist, spreading west from its original colonies in south-east England.

These two species are very similar, their size difference rather insignificant. Only the Red-eyed is on the wing in spring, and it remains more widespread (for now) than the fast-increasing Small Red-eyed.

How to find

■ **Timing** The Red-eyed appears in mid-May, and flies until early September. The Small Red-eyed has a shorter season, flying from early July to mid-September.
■ **Habitat** Both species prefer still, larger water bodies with some floating vegetation, which they use as a platform from which to lay their eggs, as well as a territorial vantage point.
■ **Search tips** Look out for these damselflies perched on lily pads or other floating plant-matter. They may be some distance from the shore, so binoculars or a camera with a long lens can be helpful, especially the latter to confirm identification.

WATCHING TIPS

Because they spend their time out on the water, these damselflies are not easy to watch, but a combination of binoculars and a small lake should produce good views and allow you to watch their interactions. Males use a lily pad or clump of floating weed as a territory, keeping other males away. Once a male and female have paired and mated, they stay joined together, the male standing on the lily pad while the female balances on the edge, curling her abdomen underneath to lay eggs on the underside of the floating leaf.

Large Red Damselfly

Pyrrhosoma nymphula

Length 2.9cm

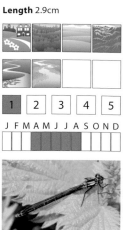

| 1 | 2 | 3 | 4 | 5 |

J F M A M J J A S O N D

This is the only red damselfly you'll see over most of Britain. It is very common and widespread, occurring throughout Britain and Ireland, and although not usually very abundant, it is usually present in small numbers at even the tiniest pond. The male has a red abdomen with black markings at the tip, and a black thorax with red top-stripes. Females are variable – they may be mostly red with black at the tail tip, mostly black with just a little red at the top of the abdomen, or intermediate. Males and all females have black legs and red eyes.

This damselfly is often seen hunting around vegetation that may be quite a distance away from the nearest fresh water. It is quick to colonise new garden ponds, provided there is some emergent vegetation.

How to find

■ **Timing** This damselfly emerges in late April, and is the first Odonata species most watchers are likely to see each year. It flies into late August.

■ **Habitat** It occurs around ponds, lakes, streams, boggy pools, canals and ditches, and immature individuals may be found hunting away from water, around woodland edges or in fields.

■ **Search tips** The red colour draws the eye as this species flies past; you can also find it resting on sunlit vegetation on warm days. However, it is rarely as active or noticeable as the blue damselflies.

WATCHING TIPS

Males hold small territories, and guard them from a perch, flying out to chase other single males, males joined to females and unaccompanied females. Several pairs may assemble together at good egg-laying spots. If you create a new garden pond, the Large Red Damselfly is likely to be the first kind of damselfly to arrive, giving you the chance to watch this small but efficient predator in action all through the summer – you could also catch its nymphs if you try out some pond-dipping.

Ceriagrion tenellum

Small Red Damselfly

Length 2.5cm

1	2	3	4	5

J F M A M J J A S O N D

It shares its colour with the Large Red Damselfly, but otherwise the Small Red Damselfly has little in common with that species and the two are not closely related. The Small Red has much more specialist habitat needs, and accordingly a more restricted distribution – it is found only in parts of southern England and western Wales, with a small outpost in northern East Anglia. The male has an entirely red abdomen and black thorax. Females vary from male-like to all black, and half red, half black, but in all cases can be distinguished from the Large Red by their red, rather than black legs.

This is a tiny, delicate insect, rather inconspicuous in its habits. Numbers can be dense in good habitat – look out for the three female colour variants.

Super sites

★ Thursley Common, Surrey
★ New Forest, Hampshire
★ Somerset Levels, Somerset

How to find

■ **Timing** It flies from early June to early September.
■ **Habitat** This is a damselfly of boggy, acidic pools and streams with marginal sedges and other vegetation, also shallow pools on chalky soils in some areas.
■ **Search tips** You may spot this very small damselfly warming up on logs or stones near the water. The bright red males may be striking in flight but the darkest females are hard to spot – be patient and check waterside vegetation carefully.

WATCHING TIPS

Males defend very small territories, which females visit from mid-morning to mid-afternoon. Paired couples spend an hour or more copulating, often well-hidden to avoid too much harassment from other males, which would try to force the pair apart. Later the pair stay in tandem while the female lays her eggs on the stems of emergent water plants, sometimes climbing down the stem so deep underwater that she submerges her mate as well as herself.

Small Red Damselfly · BRITISH NATUREFINDER **205**

Migrant dragonflies and damselflies

Ceriagrion tenellum

1	2	3	4	5

J F M A M J J A S O N D

Lesser Emperor Dragonfly

Green Darner Dragonfly

Dragonflies are highly mobile and it is no surprise that several species are recorded as rare migrants or vagrants. The list even includes a species from North America, the Green Darner, which occurred in south-west England in 1998 following hurricanes. More regular visitors include Vagrant and Lesser Emperor, the latter having bred on several occasions, and the Southern Migrant Hawker, now colonising wetlands in Essex. Several species of darter from the continent have been recorded, some just once during an exceptional influx in windy weather in 1995. Rare damselflies also appear from time to time and attempt to establish colonies – they have included Dainty Damselfly and Southern Emerald.

Finding a rare vagrant dragonfly is no less exciting than happening upon a rare bird, and will attract a high level of interest from other local wildlife-watchers. Time is of the essence if you decide to 'twitch' a dragonfly, as it may not survive very long!

How to find

■ **Timing** With January sightings of Vagrant Emperor on record, migrant Odonata could occur at any time of year, although summer or autumn are the most likely times.
■ **Habitat** Migrants are most likely to be found near the south or east coasts, and will be drawn to rich, productive wetlands.
■ **Search tips** Check any interesting dragonfly or damselfly found near the coast, and particularly try sites that have historically turned up records. Dungeness in Kent, for example, is a good place to search as it is a headland pointing towards the continent and has much suitable habitat. Social media groups can help with spreading the word about interesting sightings.

WATCHING TIPS

If word gets out that a rare dragonfly is around, interested wildlife-watchers converge at the site in the hope of seeing the insect. Find the crowd and you will hopefully soon see the insect. If viewing is difficult, it can be tempting to break the rules of wildlife-watching and trespass where you shouldn't, but you must still respect wild habitats and people's land, as well as avoid harassing the dragonfly itself. If you are patient, most Odonata will give good views sooner or later. And if you find a rare dragonfly, consider the possible impact of many visitors at the site in question before telling anyone other than your local recorder.

Apis mellifera

Honey Bee

Length 15mm

1	2	3	4	5

J F M A M J J A S O N D

Activity inside a beehive is chaotic and confusing at first glance. Watch carefully for a while though and a sense of order will start to emerge.

This is one of the most well-known and economically important insects on Earth. It produces honey which can be harvested, but more importantly it pollinates a vast range of plants, including many food crops. Many of the Honey Bees you see will 'belong' to a beekeeper, but others are wild, making their own hives in hollow trees or other sheltered places – occasionally the combs are even built in full view, hanging from a tree branch. Honey Bees can be seen everywhere in Britain, wherever there are flowering plants from which they can collect nectar and pollen.

How to find

■ **Timing** Honey Bee workers and their queen spend winter in the hive, living off stored supplies, but the workers begin to fly again in early spring. Swarms occur later in spring, when new queens are born and the old queen, together with some workers, leaves in search of a place to establish a new hive. Numbers of workers build through spring and summer, until cold weather and shortage of flowers curtails activity in mid-autumn.

■ **Habitat** Anywhere where flowers grow, you'll see Honey Bees. They visit many garden and wild flowers, particularly nectar-rich types like heather, lavender and buddleia. In early spring, wild flowers like bugle and primroses are important, while ivy flowers are a valuable food source in autumn.

■ **Search tips** Look for Honey Bees anywhere there are flowers. Swarms are often found resting on buildings or tree branches, and look at a distance like a large brown lump, as the bees are packed so tightly together.

WATCHING TIPS

You can easily watch Honey Bees gathering nectar and pollen in your garden. Their social lives are famously complex but most interactions take place within the hive. Some wildlife reserves and museums have a 'show hive' behind glass, so you can watch the bees on their combs, carrying out routine maintenance, and performing their 'waggle dances' which direct other outgoing bees to good nectar sources. A Honey Bee swarm may look alarming, but because the bees have left their hive and so have nothing to defend, they are very unlikely to sting. If a swarm settles in a problematic place, contact a beekeeper to collect it safely.

Common bumblebees

Bombus

White-tailed Bumblebee | Common Carder Bumblebees

1	2	3	4	5

J F M A M J J A S O N D

Red-tailed Bumblebee

Britain is home to a couple of dozen species of bumblebees. Over most of the country you will easily find Buff-tailed, White-tailed, Red-tailed, Common Carder, Early and Garden Bumblebees, as well as the recent colonist the Tree Bumblebee, which is spreading rapidly. They can be identified by their colours and patterns – The Bumblebee Conservation Trust provides a free online identification guide to common species (visit bumblebeeconservation.org and follow the links). Queen bumblebees, which are the only members of each nest to survive through winter, are markedly larger than workers and are often spotted very early in spring.

Male bumblebees usually have more yellow fur than the worker females do – often including an impressive blond moustache, as well as an extra band or two of yellow on the thorax or abdomen.

How to find

■ **Timing** Look out for queens as soon as the air warms up, even as early as February – they will tend to appear before the earliest spring butterflies.
■ **Habitat** Anywhere with plenty of flowers will attract bumblebees. Nest sites are often underground, in old rodent burrows – Tree Bumblebees use hollows in trees and frequently colonise bird nestboxes.
■ **Search tips** Watch any bank of flowers on a sunny day and you'll probably soon see several bumblebee species. Fields of flowering lavender, and heathlands with abundant heather, can be extremely busy. In woodlands in summer, look for large stands of bramble in flower to see plenty of bumbles.

WATCHING TIPS

These much-loved insects are charming to watch as they buzz from flower to flower and move in their classic blundering way across the blooms. Some flower types require the bee to climb inside to access pollen and nectar, such as foxgloves, and Himalayan Balsam which leaves the bees dusted white with its pollen. With some other flowers, such as comfrey whose corollas are too narrow for bumbles, you may see bees 'cheating' by making a hole in the flower at its base, to access nectar directly. If bumblebees nest in your garden, leave them to their own devices where possible – nests are usually small and will die out at the end of summer.

Rare bumblebees

Short-haired Bumblebee   Great Yellow Bumblebee

1	2	3	4	5

J F M A M J J A S O N D

Shrill Carder Bee

There are more rare or localised bumblebee species in Britain than common ones. Probably the rarest is the Short-haired Bumblebee, formerly extinct here but reintroduced in the early 2000s at Dungeness in Kent. A spectacular rarity is the Great Yellow Bumblebee, found on flowery machair grassland on northern Scottish islands. In south England, you could find the Shrill Carder Bee, which has a distinctive high-pitched buzz. Britain is also home to six species of cuckoo bees or cuckoo bumblebees, which closely resemble 'true' bumblebees (see p210). Visit the Bumblebee Conservation Trust's website (bumblebeeconservation.org) for images and distribution maps for all British species.

Bumblebees as a group are not too difficult to get to grips with. The Bumblebee Conservation Trust's website will help you discover where to go to find the various rare species, and also offers advice on attracting all bumbles, common and rare, to your garden.

How to find

■ **Timing** The life cycles of bumblebees are all alike, with queens flying first in early spring, and workers appearing some weeks later after the queens have established hives and laid eggs stored from mating the previous year. Some species mainly disappear by mid-summer, while others fly on into autumn – by the end of the year, though, all that remains are young queens, which hibernate alone in a sheltered spot.

■ **Habitat** Bumblebees of all kinds visit flowers for nectar, so may potentially be seen on sunny days, anywhere where flowers grow. Check the habitat requirements of scarcer species – for example, the Blaeberry Bumblebee is likely to be found on heather moorland.

■ **Search tips** Look out for bumblebees visiting flowers in sunny weather – activity increases as the day goes on. You may find new queens and mating pairs out in the open later in summer. On cool or rainy days you may spot them clinging to a flower-head, resting and waiting for the weather to improve.

WATCHING TIPS

Check the bumblebees in your garden – if you are in range of scarcer species, they may well visit, alongside their commoner relatives. Once in a while you'll find an exhausted bumblebee on the ground, unable to take flight. You may be able to revive it by offering it a drink of water with some sugar dissolved in it – often this is enough to restore the bee's strength. It won't work every time as worker bumbles have naturally short lifespans, but saving a weakened queen in autumn could mean there will be an extra colony the following year.

Cuckoo bees

Bombus

Vestal Cuckoo Bumblebee

J F M A M J J A S O N D

Gypsy Cuckoo Bumblebee

These bees are technically bumblebees but differ in their way of life. They do not create their own nests but invade those of other bumbles and lay their eggs there, to be unwittingly raised by the hosts. Britain has six species, of which all are fairly widespread (especially the south), but only three (Gypsy, Barbut's and Red-tailed) occur in Ireland. The Red-tailed is absent from Scotland. They are mostly mimics of other species but have dark rather than clear wings, and no pollen baskets on their hind legs. Because they don't produce their own workers, they are never as numerous as the true social bumblebees.

The cuckoo bees have very interesting life cycles, though sadly for us all of the drama takes place out of sight inside bumblebee nests. They rely on healthy populations of their host species to survive.

How to find

■ **Timing** Cuckoo bees have no queens. The adult females hibernate, and fly from spring, searching for a host nest. Then adults of both sexes appear in summer.

■ **Habitat** Like true bumblebees, cuckoo bees come to flowers to feed on nectar, but as they are only taking care of themselves they don't need to spend as much time engaged with this task. Look out for them in the vicinity of social bumblebees' nests.

■ **Search tips** Cuckoo bees are easily overlooked. Make a habit of checking bumbles for yellow pollen baskets on their hind legs – if you see no yellow, you may have found a cuckoo bee.

WATCHING TIPS

These bees have a fascinating if underhand life cycle. After sneaking into a nest and making herself at home, the female bee lays her own eggs in the nest, and may also kill the bumblebee queen. The workers are unaware that they are raising imposters in their own nest. Sadly, this drama will almost certainly go unobserved, but look out for cuckoo bees pairing and mating late in summer. If you have a bug hotel of some kind, female cuckoo bees may hibernate in it.

Mason bees

Red Mason Bee

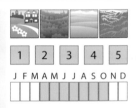

| 1 | 2 | 3 | 4 | 5 |

J F M A M J J A S O N D

Blue Mason Bee

Bees of the genus *Osmia* are popularly known as mason bees because of their liking for nesting in cracks and crevices in brickwork. The most common species in Britain is *Osmia bicornis*, often called the Red Mason Bee. It resembles a rather small, fluffy and red Honey Bee, and commonly nests in house walls, sometimes in such numbers that you can easily form the impression that it has a nest or hive within. However, this is a solitary bee and each female has just one nest chamber, separate from all others. There are several other *Osmia* species in Britain, including the widespread, blue-black *O. caerulescens*, and the rare Scottish wasp mimic *O. uncinata*. All are small bees but not as delicate as most other solitary bees.

Although they possess miniature stings, mason bees are not aggressive and would only ever sting in an extreme self-defence situation. If you have them nesting in your walls, plant nectar-bearing native flowers to keep them happy through the summer.

How to find

■ **Timing** You will see Red Mining Bees from April through to late June. Some species produce two broods a year, flying into autumn, while in others the bees may take two or more years to complete their larval growth.

■ **Habitat** These bees feed from flowers, and make nests in existing crevices and holes – often brickwork in the case of Red Mining Bees, but other species may use the old feeding galleries of beetle larvae in rotting wood, or gaps under stones.

■ **Search tips** Look out for mason bees prospecting holes in walls and visiting bug hotels in spring and early summer. Any small hole that has been blocked with a little mud plug may be an *Osmia* nest.

WATCHING TIPS

If you have buildings with old brickwork, or a bug hotel, or both, you are likely to see Red Mason Bees and possibly other *Osmia* species too. Each nest contains several brood cells in which the female lays an egg – she places eggs destined to produce male bees at the front so they emerge first the following spring. You may see her bringing pollen to the nest, then eventually carefully sealing it up with mud. Seeing these bees in profusion around your walls may cause alarm but there is no need – they are placid, and they cause no structural damage.

Andrena bees

Andrena

1	2	3	4	5

J F M A M J J A S O N D

Ashy Mining Bee (female)

Tawny Mining Bee (female)

There are about 60 British species in this genus of solitary bees. They are popularly known as mining bees, because the females dig out their own nest holes, and include some very attractive species. The Tawny Mining Bee, with its very furry bright red body, and the black-and-grey Ashy Mining Bee, are two commonly seen and distinctive species. Many others look like small, delicate Honey Bees. The group includes some rare and localised species. Most excavate their nests in patches of dry, soft ground, and so can be easily encouraged in most gardens – the usual bee-attracting flowers will draw them in to feed.

These usually rather small-sized bees are quite appealing with their often distinctive patterns and big eyes. Spend some time watching nectar-rich flowers and you should spot a few among the more obvious bumblebees.

How to find

■ **Timing** Depending on species, the flight season usually spans a couple of months or more in spring or summer. Most species are small but one or two are bigger than Honey Bees.
■ **Habitat** These bees need flowers from which to feed, and somewhere to dig their nests. Banks of sandy earth with some overhanging vegetation for shelter will attract many *Andrena* bees.
■ **Search tips** Look for these little bees on flowers through the warmest months, and for their small nest holes in the ground. Watch a likely-looking hole for a time and you may see the bee returning to it.

WATCHING TIPS

These bees are charming and present some serious identification challenges. A macro adaptor for your smartphone's camera lens will enable you to take detailed pictures, helpful for identification, and also reveal interesting anatomical detail, such as the tiny extra 'eyes' on the forehead between the large compound eyes. If you make a large bug hotel with its base on the ground, adding a section of dry earth in a space at the bottom should attract some *Andrena* bees to nest.

Other solitary bees

Eucera longicornis   Pantaloon Bee

Little Flower Bee

Besides *Osmia* and *Andrena*, there are dozens more solitary bees that occur in Britain and Ireland. More than 90% of our bee species are solitary and, while they are less showy than the industrious Honey Bee and plump buzzing bumblebees, they include some very attractive species with fascinating habits. Among them are the Leafcutter Bee with its habit of snipping circular holes in rose leaves; the rare and beautiful *Eucera longicornis*, with its oversized antennae; the Pantaloon Bee, the female of which appears to be wearing oversized fluffy yellow trousers; and the flower bees (*Anthophora*), which are bumblebee-like and include the striking stripy-bodied and green-eyed *Anthophora bimaculata*.

An interesting variety of commoner solitary bees can be attracted to the average garden – a good starting point when getting to know Britain's 250 species! To attract them, make sure you have nectar-bearing plants that between them flower from early spring to late autumn.

How to find

■ **Timing** Solitary bees of various species are on the wing from early spring until autumn. One of the first is *Anthophora plumipes*, the female of which is black and fluffy with yellow hindlegs. Late in the year you may see the Ivy Bee, a recent new colonist from the continent, which flies from September to November.

■ **Habitat** There are solitary bees in all habitats which have flowers and potential nest sites. Start with your garden and local park.

■ **Search tips** Solitary bees can be found visiting flowers of all kinds. They are usually more discreet when visiting their nests, but any small hole excavated in soft ground is worth a watch, to see if a bee emerges or arrives. Completed nests are usually closed up and hard to spot.

WATCHING TIPS

This is an excellent example of a less well-known group of species which will greatly reward your time and care in getting to know them. They are also readily encouraged into the garden. Sow plenty of nectar-rich wild flowers, covering the whole season from spring to autumn, and provide a selection of nest sites – bug hotels are good for mason bees, and others can be enticed with piles of sandy earth, arranged in a slope or hill and held in place with sections of chicken wire.

Bug hotels

There are few simpler or more satisfying things you can do to boost your garden's wildlife appeal than to install a bug hotel. These offer hiding places and nesting sites for a great variety of invertebrates, from spiders to centipedes and lacewings to ladybirds. All of these 'minibeasts' attract bigger beasts, such as songbirds, shrews, perhaps lizards and amphibians, forming a biological army against slugs, snails and other plant-munchers that you may not want to proliferate too much in your garden. Bug hotels are also, of course, hugely valuable to solitary bees, which are in turn hugely valuable to us thanks to their pollinating activities.

The structure

Choose a spot that gets plenty of sun but has some shade and shelter from the elements, ideally close to the best part of your garden for wild flowers. Large, varied bug hotels will attract more wildlife, but may not suit the look of every garden as they are usually made from natural materials and consequently have a rather rustic appearance. However, they can be attractive nonetheless – and in any case, if you have a

Bird nesting boxes are often claimed as nest-sites by queen bumblebees.

wildlife-friendly garden its whole appearance is likely to be on the rustic side. If you prefer something on the more modest side, you can buy small bug hotels, the size and shape of a bird nestbox and filled with neatly cut bamboo sections, which can be hung from a wall or placed on a pole.

A female **Blue Mason Bee** inspects a perfectly sized hole in a bug hotel.

For a larger structure, start from the ground up and build a framework of poles sunk into the ground, or stacks of bricks, perhaps held together with chicken wire, and make different 'compartments' within. You can then place different materials in each compartment, with heavier items near the bottom and lighter ones near the top. Whatever the size of your hotel, a peaked roof is advisable to encourage water to run off, and internal 'floors' should also allow for drainage. Stability is very important for large hotels, especially if you have children who might try to climb on the structure.

Bug hotels can be very simple – just a few holes drilled into a slab of wood – or real works of art. They offer gardeners the chance for creative expression while providing shelter for some of our most important and attractive wildlife.

The accommodation

A good bug hotel has a range of options for its guests. Classic components include stacks of bamboo stem sections, their open ends facing outwards, and logs or bricks with holes drilled into them. Also try bundles of twigs of various sizes, piles of large and small stones, and pine cones – any reasonably dry material that offers gaps of different sizes for shelter will be useful to invertebrates.

Bug hotels will not last forever and you should be prepared to replace and rebuild bits from time to time. Check holes and bamboo stems for occupants before throwing anything on the bonfire, and be careful not to crush any of the guests while working on the hotel.

Earth banks

An earth bank can form part of a bug hotel – just place a sloping pile of dry, sandy soil in a ground level compartment. You can also make a standalone earth bank to encourage mining bees and other soil-tunnelling insects to move in. The soil can be contained in a framework, just leave one face open, perhaps with chicken wire over it to help hold it in place without keeping out any small visitors.

Hornet

Vespa crabro

Length 3cm

1	2	3	4	5

J F M A M J J A S O N D

A fearsome-looking insect, the Hornet is our largest species of social wasp, but is of a more mellow disposition than some of its smaller relatives and rarely stings. Nevertheless, it should be treated with respect, particularly close to its nest. It has a yellow face and outer abdomen, and its eyes, thorax, legs and abdomen stripes are red-brown rather than black. It occurs quite widely in southern and central England and Wales, but is spreading northwards – it is rarely common, though. The smaller, darker Asian Hornet has recently been recorded in Britain for the first time. Report any suspected sightings to your local insect recorder.

Although Hornets are rather docile, you would still be well-advised to be very careful in the vicinity of their nests. This species rarely nests inside buildings so is unlikely to be problematic for people.

How to find

■ **Timing** Queen Hornets emerge from their hibernation in early spring, and seek a nest site. The first workers appear a few weeks later, and may be observed through summer and into early autumn.

■ **Habitat** The Hornet needs large tree hollows or similar shelters in which to nest. It hunts in all kinds of open habitats, especially open areas in or near woodland, and may visit Honey Bee hives.

■ **Search tips** Hornets draw attention on the wing because of their size and colour (though beware the Hornet Hoverfly, p225). Any habitat rich in insect life is worth searching for Hornets, and if you find good numbers you may be close to a nest.

WATCHING TIPS

This is a powerful predator, capable of catching and subduing insects as large as itself. Workers have to provision their queen and her growing larvae so need to bring home plenty of food – watch one for long enough and you may see it make a kill. If living prey is scarce, you may see Hornets raiding spiders' webs, and hunting for scraps of food around picnic tables. The Hornet is less likely to sting than the smaller social wasps, though its sudden arrival can still be quite alarming!

Bee-wolf

1	2	3	4	5

J F M A M J J A S O N D

This is a rare and remarkable solitary wasp, which superficially resembles the Common Wasp but has a dark head with just a small yellow patch on the front of its face and a fairly long, slim yellow abdomen with narrow, evenly spaced black stripes. Its legs are also yellow, while the thorax is solid black. It has a large and wide head relative to its size, wider than the rest of its body, and its wings are orange-tinted. It is found sparsely in southern and eastern England, particularly in East Anglia, and the recent discovery of several new colonies may indicate it is increasing and spreading – or that more wildlife watchers are becoming aware of its existence.

The Bee-wolf is fast-moving, both on the ground and in the air, and has a large head and strong jaws, as befits its status as an alpha predator in the insect world.

Super site

★ RSPB Minsmere, Suffolk

How to find

■ **Timing** Bee-wolves are active from early July to mid-September.

■ **Habitat** It occurs on warm dry heaths with bare ground, on sand dunes, and other similar habitats. It is expanding its range and could be found well away from currently known colonies.

■ **Search tips** Look out for it visiting flowers, and for the females digging out their nest-holes in soft earth using their powerful jaws and front legs, then carrying prey back to the nests.

WATCHING TIPS

Like many solitary bees, this species has the gruesome habit of stocking its nest with living prey which has been paralysed with a sting, to serve as fresh food for its larvae. In the case of the Bee-wolf, that prey is always the Honey Bee, and if you watch an area inhabited by Bee-wolves, you will sooner or later spot a female flying along in low direct flight, holding an upside-down Honey Bee with her legs. Back at the nest she stuffs herself and the paralysed bee into the hole and lays her eggs on its body. Each nest may contain 30 or more brood cells, each of which may be stocked with up to five Honey Bees.

Social wasps

German Wasp

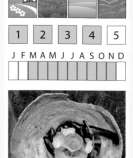

| 1 | 2 | 3 | 4 | 5 |

J F M A M J J A S O N D

Common Wasp building nest

Most of us would perhaps prefer to learn how not to be found by the two most frequently seen social wasps in Britain – the Common and German Wasps, both of which are a nuisance at picnics and will sting more readily than other bees and wasps, especially near the nest. However, they are interesting insects to watch when you are not trying to keep them away from your glass of lemonade, and there are a number of scarcer social wasps in Britain to seek out as well, such as the colourful Red Wasp, and the Saxon Wasp, a recent colonist present in southern England only.

A wasp nest is a true work of art, although quite understandably not welcome inside any building that's regularly in use, as its occupants can be fiercely protective of it.

How to find

■ **Timing** Queen wasps emerge from hibernation in early spring, and go in search of a place to nest. Later in spring, workers appear, and by late summer there may be very large numbers on the wing, though they die off through autumn.

■ **Habitat** Social wasps occur in all kinds of habitats where food may be found, and they take a wide range of animal prey as well as nectar. The nests are usually built inside hollow trees, but sometimes in buildings.

■ **Search tips** All of these wasps construct their nests from paper which they make by chewing wood to a pulp. In spring and early summer, when the first workers are busy expanding the nest, you will often find them on wooden fences and railings, chewing off wood for the nest. Later in the year, feeding the family takes all their time and they roam widely in search of food.

WATCHING TIPS

Like other social insects, these wasps show complex and interesting behaviours, and the nests they build are truly works of art. Despite this, few of us would willingly put up with an active nest in our house or even an outbuilding – unlike bumblebees, wasps will aggressively defend their nest, and once one wasp has stung, this releases a pheromone that attracts others to join in. However, wasps in the garden provide valuable pest control and pollination services.

Ruby-tailed wasps

Chrysis fulgida

Chrysis ignita

These small solitary wasps are also known as 'jewel wasps' because of their shining iridescent colours, and 'cuckoo wasps' because of their breeding behaviour. The most common British species is *Chrysis ignita*, which is widespread throughout Britain, but there are a number of others which are very similar. Despite its tiny size, *C. ignita* is very noticeable because of its bright colours – the head and thorax are metallic blue-green, the abdomen violet-red. A closer look reveals large black eyes and long antennae.

Ruby-tailed wasps' brilliant iridescent colours draw the eye, but they are often frustrating to watch as they move so quickly.

How to find

■ **Timing** *C. ignita* flies from April to September. Other ruby-tailed wasps are active at various times in spring and summer.

■ **Habitat** Wherever mason bees or other solitary bees and solitary wasps occur, you may find this species as it is a parasite that lays its eggs in their nests. It is often spotted on sun-warmed walls or fences.

■ **Search tips** Spend some time watching your bug hotel and you could spot this interloper rushing about, looking for an occupied nest. It is a very active and restless insect, quick to vanish if alarmed, so keep still and be patient. Identifying these wasps to species is not usually possible in the field.

WATCHING TIPS

There are few more dazzlingly coloured creatures in the British countryside than the tiny ruby-tailed wasps. The common *C. ignita* has no particular host preference, and you may spot it slipping into any occupied solitary bee nest that has not yet been sealed. It lays its eggs next to the host's eggs, and when the eggs hatch its larvae feed on the host larvae. Therefore, next spring it could be a ruby-tailed wasp that makes its way out of a sealed-up bee nest .

Other solitary wasps

Digger wasp

1	2	3	4	5

J F M A M J J A S O N D

Nomada wasp

Britain is home to thousands of solitary and parasitic wasp species, including cuckoo wasps that lay their eggs in the nests of social species. Many others place paralysed prey in their nests for their larvae to feed on, while others lay their eggs on the bodies of (much larger) prey and the larvae consume the living tissues without killing the host until they have finished with it. Some females of solitary wasps have startlingly long ovipositors (egg-laying tubes) – *Gasteruption jaculator*'s is longer than the rest of its body. These might be mistaken for stingers, but are used to insert eggs into hard-to-reach places. Other tiny wasps lay their eggs inside plant tissues and cause distinctive abnormal growths to form.

Solitary wasps may be confused with (true) flies. Check for eyes taller than they are wide, a narrow 'waist' where thorax meets abdomen, and obvious, often robust antennae.

How to find

■ **Timing** Solitary wasps may be found any time between spring and autumn, depending on when their preferred prey or host species is available – a few species are active in winter.

■ **Habitat** Most parasitic wasps concentrate on one or a few species, and occur wherever the host or prey is to be found. Those that dig nests also need places with soft bare ground.

■ **Search tips** These mostly small wasps can be difficult to find, but evidence of their activities may be more obvious. Oak trees commonly have several kinds of galls – spangle and marble galls on the leaves, 'oak apples' on the stems, and knopper galls which are deformed acorns – each is formed by a different kind of wasp. It is common to find dead Large White butterfly caterpillars surrounded by yellow cocoons of the wasp *Cotesia glomerata*.

WATCHING TIPS

With luck and patience, you stand a good chance of catching some of the solitary wasps in action. If you often raise caterpillars, you will probably sometimes encounter solitary wasps emerging from pupae rather than the expected moth or butterfly. Almost every kind of invertebrate has its own parasitic wasp to contend with, sometimes several, specialising in different life stages. Identifying these wasps can be extremely difficult, even with clear photographs, but if you also observe the host species, that can help narrow down the possibilities.

Urocerus gigas

Giant Wood Wasp

Length 4cm

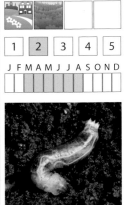

1	2	3	4	5

J F M A M J J A S O N D

Giant Wood Wasp larvae cause extensive and obvious damage to wood during the several years it takes them to grow to maturity. Their tunnels form a dramatic pattern when the wood is cut open.

This spectacular insect is a sawfly, rather than a wasp, and is also known as the Horntail. It has a yellow abdomen with a broad black band across the middle, a black thorax, a black head with yellow markings behind the eyes, and yellow antennae and legs. The female has a long, strong ovipositor, more than half as long as her abdomen, giving her an alarming appearance, but this species is quite harmless – like all sawflies, although it is a relative of bees and wasps, it does not possess a sting. It is widespread and fairly common in Britain.

How to find

■ **Timing** This insect may be seen any time between May and August.

■ **Habitat** Giant Wood Wasps are associated with pine trees, and so are most likely to be found in pine forests, but will also turn up in gardens and parkland, and at plantation edges where piles of pine logs are stored.

■ **Search tips** Look out for this species settling on partly decaying pine trees or on pine logs or planks. It may attempt to lay eggs on – or even emerge from – pine-framed outbuildings or outdoor furniture. It sometimes enters buildings – catch it gently with a glass and piece of card, and release it away from the house.

WATCHING TIPS

This sawfly lays its eggs within pine wood, the female pushing her ovipositor as deeply as she can into the wood and injecting her eggs. The larvae feed on the wood, making extensive tunnel systems over their long lives (this species may take five years to mature and pupate). Look out for female Giant Wood Wasps prospecting for egg-laying sites at the sunny edges of pine forests, as well as flying around logpiles in sunny clearings.

Other sawflies

Rose Sawfly

Turnip Sawfly

Sawflies are the often overlooked 'fourth member' of the order Hymenoptera, alongside bees, wasps and ants. Nearly 550 species occur in Britain, some very common but many others localised or very rare. Adult sawflies have two pairs of wings, and usually lack the distinct narrow waist of wasps and bees, where the thorax and abdomen join. They lay their eggs on plants, and often their larvae bear a striking resemblance to Lepidoptera caterpillars – however, Lepidoptera always have five pairs of prolegs at the rear of the body, and sawflies always have more than five. Sawfly larvae also have a habit of raising their rear ends into the air while they feed.

Sawflies are often quite colourful, and have a more robust anatomy than small solitary bees and wasps. They do share with them the traits of elongated eyes and prominent antennae, which help distinguish them from true flies.

How to find

■ **Timing** Different types of sawflies appear at different times through spring, summer and autumn. High summer is the time for highest diversity.
■ **Habitat** These insects' distribution is linked to that of their foodplants. Several species, both common and rare, feed on garden plants.
■ **Search tips** Adult sawflies feed on pollen and nectar, so may be seen on flowers of all kinds – look for them on sunny summer days in particular. They tend to be plump-bodied and slow-moving. The larvae of some species are quite obvious as they feed in groups on leaves, working their way through each leaf in parallel formation.

WATCHING TIPS

This insect group is not very well known, but there is every chance that a good variety occurs in your garden or local park, including distinctive and attractive species like the Large Rose Sawfly with its bright orange body, the pretty green and black *Tenthredo mesomela*, and the beautiful shining gold Honeysuckle Sawfly. The gregarious larvae of particular species can be somewhat damaging in the garden, though their presence will attract predatory insects and birds.

Formica rufa

Wood Ant

Length 10mm

1	2	3	4	5

J F M A M J J A S O N D

This very large ant is responsible for the large mound-shaped nests that you'll find in woodland through much of Britain. Those in Scotland belong to two different species, *F. aquilonia* and *F. lugubris*, but all three are very similar in appearance and habits. A good-sized nest may stand a metre high or more, and be home to more than 100,000 worker ants. The Wood Ant has a black-brown head and abdomen and a red-brown thorax. Workers are wingless but newly emerged queens and males have two fragile pairs of wings. Queen Wood Ants sometimes take over nests of other, smaller ant species.

A well-established Wood Ant nest is easily spotted, thanks to its huge size. Check the ground and the trunks of trees nearby and you'll see the brown-and-black worker ants coming and going.

How to find

■ **Timing** Wood Ants will be seen around the nest throughout the warmer months. Through winter, activity continues inside the nest but the ants do not venture outside.

■ **Habitat** This is a species of all kinds of woodlands. Nests are usually constructed in sunny clearings and often pressed against a decaying tree stump. It is not unusual for several nearby nest mounds to be connected underground.

■ **Search tips** Once you have found the conspicuous nest mounds, you will find the ants, walking along scent trails left by their fellows and searching for food – anything edible that they can physically carry back to the nest. Up in the trees you may see them tending their 'herds' of aphids, which they guard and milk for honeydew.

WATCHING TIPS

Ants have fascinating social behaviour, although it takes time and care to observe and interpret their activities. Ant watching is also slightly risky – Wood Ants will bite, and then squirt stinging formic acid on the tiny wound they have made. Visit nests in spring for a chance of seeing queens and males in their mating flights. Queen Wood Ants can live for 20 years, and a nest may contain several, so nests can exist for many years.

Other ants

Black Garden Ant

| 1 | 2 | 3 | 4 | 5 |

J F M A M J J A S O N D

Yellow Meadow Ants

Globally, ants are extraordinarily abundant – their combined weight would easily outweigh the world's seven billion human beings. In Britain, we have some 65 ant species, ranging from the familiar Black Garden Ant that nests under our patios, to the Thief Ant, a tiny species which survives by stealing food from other ants' nests, and the rather less sneaky blood-red Slave-maker Ant, which takes over other red ants' nests and steals larvae and pupae from other nests to use as slave labour. As well as our many diverse ant species, be aware of various convincing ant mimics, such as the Velvet Ant, a fluffy-bodied wingless wasp, and the Ant Damsel Bug, which closely resembles a black ant in its immature form.

There is power in numbers, and watching ants' social interactions and cooperative behaviour reveals how much we have yet to understand about these amazing insects and their complex, many-layered societies.

How to find

■ **Timing** Ants are active all year, though you are more likely to see them out and about in the warmer months.
■ **Habitat** They inhabit all kinds of countryside, as long as there is a food supply and somewhere to make a nest.
■ **Search tips** Ants are easy to find once you get down low and look for them moving through the grass. Aggregations of aphids are often attended by ants. Some species, such as the Yellow Meadow Ant, make noticeable anthills, which appear as large bumps in open grassland. The presence of Green Woodpeckers is characteristic of habitats that hold plenty of Yellow Meadow Ant colonies.

WATCHING TIPS

It is rather frustrating that the most interesting ant behaviour takes places deep within the nest, but watch patiently and you may see ants milking their aphid livestock, teaming up to carry large prey items to the nest, disposing of their dead colleagues' bodies or other activities crucial to keep the nest running smoothly and everyone fed. Some species are very visible at mating time, males and winged virgin queens emerging synchronously on a warm day for a frenzy of activity, after which the males die, and the queens remove their wings and begin to dig themselves a new nest.

Volucella hoverflies

Volucella zonaria

Length 18mm

| 1 | 2 | 3 | 4 | 5 |

J F M A M J J A S O N D

V. pellucens

In Britain, the genus *Volucella* contains five particularly large and impressive hoverflies, four of which are common and widespread, especially in the South. Two are mimics of hornets or large wasps (*V. zonaria* and *V. inanis*), two others (the very common *V. pellucens* and the rare *V. inflata*) are black with a broad white stripe across the top of the abdomen, which is translucent against the light, and the third (*V. bombylans*) is a furry bumblebee mimic, occurring in two colour forms that mimic Red-tailed and White-tailed Bumblebees. All may be seen visiting flowers for nectar or hovering high in woodland clearings. Like all true flies, they have just one pair of wings, the other pair being reduced to small, knobby structures called halteres.

The large semi-circular eyes and lack of obvious antennae help distinguish these hoverflies from the bees and hornets that they mimic. When seen backlit, the pale mid-section of *V. pellucens* is transparent.

How to find

■ **Timing** These hoverflies are best looked for on still, warm summer's days, and are most likely to fly towards the middle of the day.

■ **Habitat** All of the *Volucella* species occur in woodland, especially wide sunny clearings and flowery fields at woodland edges, and may turn up in larger gardens.

■ **Search tips** When hovering high, these insects are eye-catching. When they visit flowers they may be confused with the species they mimic but check the head structure carefully to identify them.

WATCHING TIPS

These are among the most spectacular British flies, with the hornet-mimicking *Volucella zonaria* particularly striking and causing many an unwary wildlife-watcher to retreat in alarm. They also have interesting life-cycles, the females laying their eggs inside social wasps' nests. The larvae develop within the nest, climbing over the paper cells of the wasp nest eating organic debris, but also some wasp eggs and larvae. Discovering the flattened, rather woodlouse-like larvae in your home is a sign that there's a wasp nest in the loft.

Other hoverflies

Marmalade hoverfly
Marmalade hoverfly

| 1 | 2 | 3 | 4 | 5 |

J F M A M J J A S O N D

Brindled hoverfly

Britain is home to at least 270 species of hoverflies, making this a challenging group for the amateur wildlife watcher. Recognising any given insect as a hoverfly is not so difficult though – these flies have a distinct, agile flight, interspersing hovers on the spot with quick short dashes. Eventually they make a precision landing and usually fold the wings over the body after a moment. Many larger hoverflies have striped bodies and somewhat resemble wasps or bees, which presumably discourages predators from attacking them. As with other large groups of species, get to know the common ones first, such as the Marmalade Fly with its double stripes, or the Drone Fly, a Honey-bee mimic.

As with all large wildlife groups, it's best to get familiar with the common, 'everyday' species first because this will help you to notice when you do encounter something out of the ordinary. Start in your garden or local park, and take clear photos of each species you see.

How to find

■ **Timing** Hoverflies may be seen all through the spring and summer.
■ **Habitat** These are flower-feeders, so will be found anywhere that flowers grow. They prefer flowers with wide, flat structures – few hoverflies are adapted to take nectar from flowers with long corollas and complex structures.
■ **Search tips** The tall wide flower heads of cow parsley and similar plants are good places to search for hoverflies. Also look for them hanging in the air in sunbeams within woodland clearings.

WATCHING TIPS

Clear close-up photographs are very helpful when it comes to hoverfly identification. Try to catch the insect just as it lands, before it closes its wings and conceals the abdomen pattern. Attempting to photograph hoverflies in flight is a diverting challenge for anyone with a reasonably good camera, and can also produce images that are helpful for identification. Start by exploring the species that visit your garden flowers, then have a go at recording other species further afield.

Tipulidae

Craneflies

Tipula paludosa

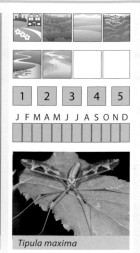

1	2	3	4	5

J F M A M J J A S O N D

Tipula maxima

These true flies are named for their exaggeratedly long, sprawling legs. They also have long, slim bodies, narrow wings and tiny heads. They are rather unpopular, mainly due to the species *Tipula paludosa*, whose 'leatherjacket' larvae damage lawns and crops, and then emerge in large numbers in late summer, swarming across fields and entering houses through lit open windows. However, there are many other cranefly species in Britain, including attractively marked species and some national rarities. Craneflies may look alarming with their huge leg-span, but they are completely harmless.

Many craneflies are near impossible to identify, but the large, boldly patterned *Tipula maxima* is one of the more distinctive. Others are less striking but may reveal interesting features on close inspection, such as the bright green eyes of *Tipula vernalis*.

How to find

■ **Timing** Craneflies of various species can be seen from spring through to autumn or winter.

■ **Habitat** These insects can generally be found in open and well-vegetated habitats, such as meadows, riversides, heaths and woodland edges. Some lay their eggs in water and so are associated with ponds and lakes.

■ **Search tips** Craneflies will often fly up from the grass as you walk along, perhaps giving the impression of a damselfly at first glance with their slim wings and body, but their weak flight and long trailing legs soon reveal their real identity. They may also visit flowers for nectar.

WATCHING TIPS

From the very large *Tipula maxima* with its attractively mottled wings to the remarkable and rare wasp-like *Ctenophora flaveolata*, the cranefly family has much to offer the interested wildlife-watcher, even though many people simply can't get past those spindly legs and will always find them unappealing. Good clear photos and expert help will be needed to identify many of the lesser-known species, but craneflies present yet another excellent opportunity for amateurs to make valuable contributions to our knowledge of wildlife abundance and distribution.

Robberflies

Asilidae family

Machimus cingulatus

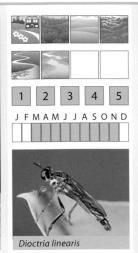

1	2	3	4	5

J F M A M J J A S O N D

Dioctria linearis

These are mostly rather large flies, with robust thoraxes giving a rather hump-backed look, longish abdomens and legs, and a slightly sinister general appearance, befitting their lifestyle as fierce predators. There are some 30 species in Britain, some common and widespread, some rare. The most impressive is probably the Hornet Robberfly, which is about 3cm long and, with its yellow abdomen, does resemble a Hornet. It is somewhat rare with scattered populations in southern England and south Wales. The Violet Black-legged Robberfly is another distinctive species with a shiny blue-black body and smoky-grey wings – it is common in southern England.

Robberflies watch for prey from raised perches. They are very fast in flight and can bring down other flying insects as heavy as themselves. Like all flies, their mouthparts are adapted to pierce and suck, so the prey is literally drained dry.

How to find

■ **Timing** Robberflies may be seen from mid-spring to early autumn.

■ **Habitat** They occur in all kinds of habitats that support plenty of prey, in the form of other flying insects. You may well find some in your garden – they will also be common in lush riverside meadows and woodland edges.

■ **Search tips** You'll often notice a robberfly when it has landed on a leaf with a prey item in its grasp. When hunting, robberflies watch for prey from a perch that gives good all-round visibility, standing very still on their long legs.

WATCHING TIPS

You need a large slice of luck and quick reflexes to witness a robberfly actually taking its prey. You are more likely to be able to see it (slowly) comsuming the prey – not so pleasant, but it will give you the chance to identify both hunter and victim – two for the price of one. Robberflies seem mostly to capture other species of fly, especially hoverflies, indicating that they are, despite their rather immobile manner when perched, very quick and agile on the wing.

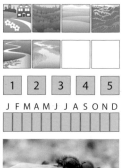

1	2	3	4	5

J F M A M J J A S O N D

St Mark's fly (Bibio marci)

Dark-bordered Bee-fly

The order Diptera means 'two-winged', and describes all true flies, in their great diversity – some 7,000 species are known from the UK alone. Besides those already discussed, the group includes the fluffy bee-flies with their sword-like mouthparts, iridescent soldier-flies, predatory snipe flies, tiny and intricately marked picture-winged and fruit flies, the various large horseflies and deerflies with their beautifully patterned eyes and powerful blood-sucking bite, and – even less popular – the houseflies, gnats, midges and mosquitoes. You will find true flies of all kinds wherever you look in the countryside and in the garden.

True flies are a very varied bunch. Some are predators, some scavengers, several are nectar-drinking pollinators, and others don't feed at all in their adult form. Those that lay eggs on carrion and animal droppings are part of nature's unsung but vital 'clean-up crew'.

How to find

■ **Timing** Flies of various kinds can be seen all year round.

■ **Habitat** The best habitats are those that are well-vegetated, but all habitats, even beach strandlines and bleak mountainsides, support some species.

■ **Search tips** Searching for interesting true flies involves nothing more than slow and careful examination of plant foliage at all levels, tree trunks and flowers. As ever, building familiarity with common species is the first step to recognising and identifying more unusual ones. Talk to other Diptera enthusiasts on social media to find out where and how to find any particular species you are keen to see.

WATCHING TIPS

In all their diverse ways of life, flies are very interesting, though not always very appealing. Wildlife-watchers always enjoy finding the first Dark-bordered Bee-fly of spring as it hovers around primroses in the flower bed. But no-one enjoys being bitten by mosquitoes or horseflies. Nevertheless, the true flies offer a vast and rewarding field of study. The Dipterists Forum (dipteristsforum.org.uk) is a valuable resource for anyone wanting help with fly identification.

Stag Beetle

Lucanus cervus

Length 5cm

1	2	3	4	5

J F M A M J J A S O N D

Male

Female

The largest British beetle, this is a magnificent animal and much sought-after by wildlife-watchers. The male is unmistakeable, with his outsized red jaws that resemble deer antlers, used for doing battle with rival males. The female lacks enlarged jaws but is otherwise similar, with red-brown wing-cases and a black thorax and head. She may be confused with the Lesser Stag Beetle, which is smaller, blacker and more 'boxy' in shape. Stag beetles are relatively widespread but rarely particularly common in southern England – however, the large parks of Greater London are particularly good for them. They also occur in the Severn valley and some coastal regions in south-west England.

Stag Beetles look imposing but are harmless – if you find one on a path or road, move it to safety.

Super sites

* ★ Richmond Park, Greater London
* ★ Bushy Park, Greater London

How to find

■ **Timing** Stag Beetles emerge from mid-May, and can be seen until late August.

■ **Habitat** This is mainly a species of woodlands, where it relies on there being plenty of decaying wood for its larvae to feed on. It is also found in parkland, and may turn up in suburban gardens.

■ **Search tips** You are most likely to meet a male Stag Beetle on a warm summer's evening, when males fly far and wide in search of females. Many Londoners have seen their first Stag Beetle fly past while they were out in the garden enjoying an evening barbecue. In the daytime, look for males basking in the sun, and females close to woodpiles.

WATCHING TIPS

The male Stag Beetle in flight looks disconcertingly huge and makes a noise to match, a loud buzzing hum. The flight is strong but a little clumsy. Once he locates a female and they mate, she makes her way to the same place where she emerged, to lay eggs in the rotting wood. Stag Beetle larvae are sometimes found when clearing up old woodpiles in the garden – if you uncover one, leave it be and replace the pieces of wood. You may have a long wait to see the adult, as it can take them seven years to mature, but this uncommon and spectacular species should be cherished wherever it occurs.

Tiger beetles

1	2	3	4	5

J F M A M J J A S O N D

Green Tiger Beetle

Heath Tiger Beetle

These fast-moving leggy beetles are dashing, conspicuous predators, and Britain is home to five species. All are rather uncommon and localised except for the Green Tiger Beetle, which also happens to be one of the largest and the most colourful of the group. It is widespread throughout Britain, and is easily recognised by its shiny metallic green coloration with a few yellow spots on the wing-cases, a broad head with large forward-angled eyes, and large jaws. The other tiger beetles have a similar structure but are duller-coloured. The biggest species is the rare Heath Tiger Beetle, which is shiny blue-black with white spots, and confined to heathlands in Dorset and Surrey.

With their very long legs, tiger beetles can sprint along at an impressive rate and don't hesitate to pounce on prey. You'll often meet them on paths that criss-cross heath or moorland, and with care you can encourage them to run onto your hand for a closer look.

How to find

■ **Timing** These beetles can be found from mid-spring to late summer or early autumn. They are active by day and prefer sunny weather.
■ **Habitat** Tiger beetles favour open countryside with bare, dry ground.
■ **Search tips** Look out for tiger beetles rushing at high speed across patches of bare ground, or making short low-level flights. You could spot the larvae's burrows in the sand.

WATCHING TIPS

These beetles are well-named, being fierce and fearless hunters. They move so quickly on foot that they are difficult to follow but if you are very lucky you could see one seizing its prey. The larvae are also predatory, digging deep burrows in bare ground and waiting for small prey to fall in – they then grab the victim with their strong jaws. Despite this, Green Tiger Beetle larvae are attacked by a parasitic wasp, *Methoca ichneumonoides*. Look out for the wingless, ant-like females investigating the beetle larvae's burrows.

Other ground beetles

1	2	3	4	5

J F M A M J J A S O N D

Violet Ground Beetle

Sexton Beetle

Beetles are by far the most diverse insect order in the world. Many are true ground beetles (family Carabidae), and some other groups like carrion and rove beetles also spend most of their adult lives at ground level. Most are predators or scavengers. Some are distinctive, such as the Violet Ground Beetle, black and glossy with shining purple wing-case edges, or the red-and-black Sexton Beetle, a carrion-eater that is often festooned with tiny pink mites. Rove beetles are distinctive with their short wing-cases and exposed abdomens. Many others lack noticeable features and have no common names – this is the domain of the serious beetle enthusiast.

Most ground beetles have fully developed wings under their elytra (wing-cases) and can fly, but they are not agile in the air and spend most of their time on the run instead. You'll often see small pink mites on Sexton Beetles – they are not parasitic but hitching a lift from one carrion meal to the next.

How to find

■ **Timing** Ground beetles may be encountered at any time of year, though summer is the best season to see a good variety of them.

■ **Habitat** You can find ground-dwelling beetles in all kinds of habitats, but particularly woodlands and grassland.

■ **Search tips** Sometimes you'll spot a ground beetle making its way purposefully along beside a pathway. A more guaranteed way to see them is to use a pitfall trap. If you find a dead bird or small mammal, you may notice carrion and burying beetles nearby. The mites carried on Sexton Beetles will eat fly larvae on carrion, leaving more for the beetles that brought them there.

WATCHING TIPS

This tricky collection of insects presents the amateur wildlife-watcher with considerable identification challenges, and they are too fast-moving to be easily photographable. If you confine your beetle to a white container for a short while, it should eventually pause long enough for useable images to be taken, which may allow you to identify it, but in some cases you will only be able to narrow it down to a genus. There are likely to be a fair few species in even a small, urban garden, while others are rare and have specialised habitat needs.

Coccinellidae

Ladybirds

1	2	3	4	5

J F M A M J J A S O N D

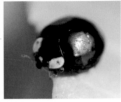

Harlequin Ladybirds are extremely variable in colour and number of spots, so can be confused with rarer species. In all colour forms, brown legs are characteristic. They will enter houses in autumn to hibernate, sometimes in large groups.

These colourful, very rounded beetles are predators that stalk the leaves and flowers of all kinds of plants, in search of aphids. Their larval stages are equally predatory, and fairly distinctive, with long dark bristly bodies and well-developed legs. There are 26 distinctive ladybird species in Britain, which can be told apart by their size, colour, and number and position of spots or other markings. Some are extremely common, including the invasive Harlequin Ladybird which now accounts for a large proportion of all ladybird sightings in southern England. Several others are highly localised. As a group, the species are few enough and distinct enough that it is quite easy for an amateur to get to grips with them.

How to find

■ **Timing** Ladybirds overwinter as adults, so may be seen from early spring when they emerge from hibernation, through to autumn when the new generation hibernates.

■ **Habitat** Some ladybirds only occur in certain habitats, for example the Heather Ladybird which is found on heathland, and the Water Ladybird which occurs in reedbeds. Most commoner species can turn up in woodlands, meadows, parks and gardens.

■ **Search tips** These beetles show warning colours to indicate to predators that they are poisonous, and this allows them to hunt out in the open. You will easily find them on plant foliage, especially in the vicinity of aphids.

WATCHING TIPS

Ladybirds are highly visible and easy to watch as they go about their business. Adults are long-lived and spend much time hunting prey prior to entering hibernation. When they emerge, they seek a mate and the female lays her eggs. Ladybird larvae and then pupae can be found through early summer. Fresh adults are pale and unmarked but soon develop their characteristic colours and patterns. Ladybird larvae are also distinctive with their long, spiky bodies, and are just as predatory as adults – they will even devour other, smaller ladybird larvae.

Watching minibeasts

The term 'minibeast' encompasses all small and usually non-flying invertebrates, including insects but also centipedes, millipedes, spiders and more. These little creatures are all around us, but we don't tend to notice them in our day-to-day lives, or even on our wildlife walks because they are often very small, at our feet or hidden in foliage, and we don't spend much – if any – time looking down or peering into bushes. To properly explore the world of small invertebrates means shifting the way you look at the world, and using a few special techniques and bits of kit.

Go low

Just lying down in a field rather than sitting in it gives you a view into the ground-level world. The longer you wait, the more you'll see, and the more you'll feel immersed in the miniature world these animals inhabit. Every habitat in every part of the countryside will offer a different experience – just make sure you're not going to lie on anything spiky or unpleasant first.

Light famously attracts moths, but beetles and other insects may come to **light traps** too.

Traps of various kinds

The pitfall trap is a classic for anyone studying small invertebrates. At its simplest it is just a container sunk into the ground, so its top is level with the surrounding earth. Some small invertebrates will fall into the open 'pit', although others may see and avoid it. Those that fall will be unable to climb out and you can have a look at them before you let them go. In rainy conditions, it is a good idea to rig up a kind of roof some centimetres above the trap (supported on corner posts, so that water does not collect at the bottom). You can also try baiting the trap, perhaps with a bit of meat or fruit.

Pitfall traps should be checked often. There is a small chance of catching something larger than you intended, such as a shrew or toad, and these should be gently and carefully released immediately.

Another method of catching small invertebrates is 'beating'. This is less violent than it sounds, and involves spreading a large sheet under a bush or other foliage, and then using a stick to gently shake the plant. Any

The **pooter** is a useful and pleasingly simple way to capture small invertebrates.

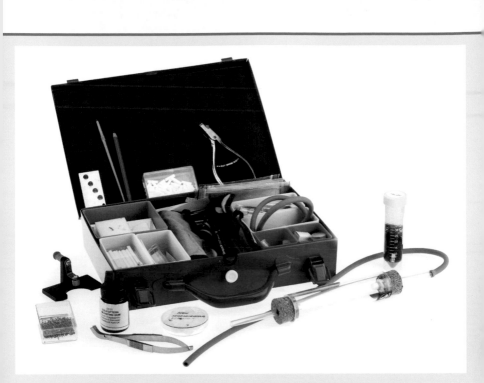

invertebrates that fall out will land on the sheet and can be examined there. You can also try using a soft sweep net to brush through long grass and potentially catch a range of small insects and spiders – have collection pots to hand to separate out your catches.

Another classic entomology gadget is the pooter. This is a pot with two tubes pushed into its lid, one with its bottom end covered in mesh. When you find an insect or other small creature that you want to catch, you place the open tube over it, and suck air through the other tube – the mesh stops you from inhaling the insect and instead it ends up safely in the pot, to be transferred to another container when you're ready. You can buy pooters from suppliers like NHBS (nhbs.com/equipment/entomology) or make your own.

A closer look

Having gone to the trouble of catching some small creatures the next step is to take a look at them, and hopefully identify them. A selection of white tubs and pots, or clear tubes, are helpful for this purpose. You can use a hand lens to examine the details

A pot with a magnifying lens for its lid makes it easy to examine your catch in detail.

An entomologist's toolkit – everything you need (and a few things you probably don't) to start studying 'minibeasts'.

close-up, or use a container with a magnifying glass lid. It is not always possible to take useful photos of tiny creatures without very specialist kit, but instead, you can draw simple diagrams of key parts of their anatomy. Purchasing a full library of field guides for all groups of invertebrates is an investment that not everyone can afford to make – an alternative is to use online resources such as Facebook and Flickr groups dedicated to the study of the organisms in question.

Great Diving Beetle and other water beetles

Dytiscus marginalis and others

Length 3.2cm (Great Diving)

1	2	3	4	5

J F M A M J J A S O N D

Great Diving Beetle

Whirligig Beetle

The Great Diving Beetle is a large, silvery beetle that is entirely at ease in the water, and it hunts quite large aquatic prey, even tadpoles and small fish. Its larva, too, is a fierce and energetic predator. This species is very widespread and common throughout Britain and Ireland. Other water beetles found in Britain include the common Whirligig Beetle, which is gregarious and spins in constant dizzying circles on the surface, and the Common Diving Beetle, an abundant black swimming beetle with a streamlined oval shape. There are about 250 more water beetles known to live in Britain.

Water beetles are adventurous and strong fliers. They are likely to be the first animals to colonise a new wildlife pond. If you dig your pond in summer you may attract them within days of the pond being filled.

How to find

■ **Timing** Water beetles are easiest to see in summer, although you may catch them or their larvae when pond-dipping at other times of year.

■ **Habitat** Freshwater ponds, lakes and rivers can all support a variety of water beetle species.

■ **Search tips** If you watch the water you may see diving beetles come to the surface from time to time to collect more air – they breathe from air bubbles they trap under their wing-cases. Some other species spend all or most of their time at the surface. You might also find water beetles away from water. Most are strong fliers and disperse in search of new waters when necessary, but can be fooled by reflective surfaces such as a wet road.

WATCHING TIPS

Water beetles should move into a new wildlife pond fairly quickly, and you can then watch them in action – a most interesting experience. Many groups of insects have an aquatic larval stage, but very few retain swimming skills into adulthood. Great Diving Beetle larvae are active underwater swimmers, with long flexible bodies. They look slightly like dragonfly nymphs but have a different head shape and spot very large scythe-shaped jaws.

Wasp Beetle and other longhorns

Length 16mm (Wasp Beetle)

1	2	3	4	5

J	F	M	A	M	J	J	A	S	O	N	D

Wasp Beetle

Dusky Longhorn Beetle

The Wasp Beetle is one of the wide variety of insects that has evolved to resemble a wasp, allowing it to feed confidently on flower heads in full view. The illusion may be less accurate when the beetle flies as its yellow-and-black striped wing-cases open up, but it does fly in a jerky, unpredictable manner, similar to a wasp. There are more than 60 other longhorn flower-feeding beetles in Britain, all characterised by long, outward-curving antennae and elongated, flat bodies. They include the attractive Golden-bloomed Grey Longhorn, which has prominent black-and-white banded antennae and dusty bronze-coloured wing-cases, and the Four-banded Longhorn, which has eight yellow or orange spots on its black wing-cases.

Longhorns tend to be elegant beetles with long bodies and long legs to go with their long antennae, and while some are a little drab there are several colourful and handsomely patterned species too.

How to find

■ **Timing** These beetles are most often seen in early to mid-summer.
■ **Habitat** Because most species feed from flowers and lay their eggs on bare, sometimes decaying wood, they are most likely to be found in flowery glades, clearings and edges of mature woodlands. They may be attracted to woodpiles in the garden.
■ **Search tips** These beetles are usually quite noticeable as they feed on flower heads, especially the broad heads of angelica and other kinds of umbellifers. They are agile in flight for beetles, and resemble bees or flies.

WATCHING TIPS

The Wasp Beetle is a distinctive longhorn, but most other species are also quite easily identifiable. They are usually easy to approach and watch, and large enough that obtaining reference photos for identification is not difficult. Submit records of interesting longhorn beetles to the National Longhorn Beetle Recording Scheme via iRecord (brc.ac.uk/irecord, or download the app).

Rose Chafer and other chafers

Cetonia aurata and others

Length 17mm

1	2	3	4	5

J F M A M J J A S O N D

Rose Chafer

Cockchafer

The Rose Chafer is a large and sturdy, rather square-bodied beetle with gleaming bright green wing-cases, marked with a few fine, wavy white stripes. The head and thorax is also shining green but a slightly yellower shade. In flight it looks like a very big, green bee. It is widespread in England and Wales, but not very common. Other British members of the chafer group (family Scarabaeidae) include the Cockchafer or Maybug, a large brown beetle with curious fan-tipped antennae, that is attracted to light in May, and the Garden Chafer, a small chafer with brown wing-cases and a green thorax and head.

Chafers are very robust beetles, which recall big, noisy bumblebees in their heavy, direct and powerful flight. A Cockchafer can cause alarm when it blunders into a room and flies noisily around the light fitting, but for all its noise and bulk it is completely harmless.

How to find

■ **Timing** Rose Chafers may be seen from May to September. Most other chafers have shorter flight seasons.
■ **Habitat** The Rose Chafer is, as its name suggests, attracted to roses (both wild dog roses and cultivated kinds). Most other chafers are foliage-eaters and can be found in various well-vegetated habitats, including gardens.
■ **Search tips** Look for the Rose Chafer visiting rose gardens in parks, and possibly garden roses too. The Cockchafer often flies to lit windows or into rooms after dark – it has an alarmingly loud buzzing flight. Seen close-up, though, it is quite appealing with a moth-like fluffy body and its unusual antennae that resemble a pair of raised hands (with seven 'fingers' in males, six in females).

WATCHING TIPS

These beetles are strong, noisy fliers and this, along with the large size of some species, can make them unpopular, along with the fact that both larvae and adults can damage crops. However, pesticide use has lowered numbers in many areas – the Cockchafer has declined considerably in parts of Britain and come close to extinction in parts of Europe. Keep your garden a pesticide-free zone if you would like the chance to enjoy more encounters with these characterful, bumbling beetles.

Flower and leaf beetles

Cantharis pellucida

J F M A M J J A S O N D

Red Soldier Beetle

Many attractive beetles can be found feeding on flowers and foliage, in early summer particularly. They include the longhorn beetles (see p237), and pretty species like the Malachite Beetle with its shining green, red-tipped wing-cases. Some other foliage-dwelling beetles are hunters, among them the Red-headed and Black-headed Cardinal Beetles, which can be seen roaming across bramble leaves and other hedgerow plants in June. Many weevils live in plant foliage too, such as the blue-green Nettle Weevil. All weevils can be identified by their long beak-like mouthparts, giving them a comical appearance. A rummage through a summer hedgerow can turn up an impressive variety of beetles.

Red Soldier Beetles are sometimes nicknamed 'bonking beetles', because in high summer hogweed and other large, open flower heads are often festooned with mating pairs. When not copulating, they are important pollinators for these kinds of flowers.

How to find

■ **Timing** Flower and leaf beetles are only around in their adult form when there are flowers and leaves to feed on or hide among, so are seen from mid-spring to early autumn.

■ **Habitat** Woodlands, mature flowery meadows, riversides, gardens and other well-vegetated habitats with flowers are good places to look for these beetles.

■ **Search tips** Check flower-heads, especially large flat ones such as Ox-eye Daisies and umbellifers – any that are attracting bees and hoverflies will also probably draw in some flower beetles. Some leaf beetles will be evident on the upper surfaces of leaves – shyer species may be found through 'beating' (see p234).

WATCHING TIPS

The range of colours, textures and body shapes in this group of beetles is impressive. Take a close look at those you find to appreciate their delicate beauty. Growing a variety of native wild flowers and herbaceous plants in your garden will help to encourage these species. A particularly good variety is active in late spring and early summer, including the common and colourful Malachite Beetle, the cardinal beetles, and the glossy green Thick-legged Flower Beetle, in which the males have bulging, almost spherical upper hindlegs.

Glow-worm

Lampyris noctiluca

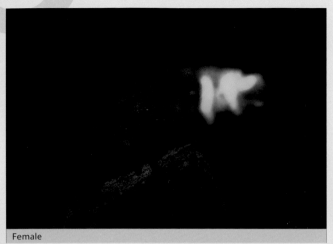

Female

Length 18mm

| 1 | 2 | **3** | 4 | 5 |

J F M A M J J A S O N D

The female Glow-worm does look like a segmented worm, with only the male looking properly beetle-like with full-sized wing-cases. However, it is the female that produces the famous eerie green glow, a signal to attract the flying males to her position. The Glow-worm larva looks much like the adult female, but emits only a very slight glow. This fascinating species is found sparsely in parts of England (especially the South), and in lowland Scotland and Wales, but not Ireland. It may be found in rural gardens, though is more typical of open countryside.

Male

Glow-worms have a similar breeding strategy to certain moths, in which females are wingless and use pheromones to attract males. In the case of Glow-worms, though, the signal is light, not scent.

How to find

■ **Timing** Glow-worms can be seen from the start of June to late August, with the females glowing at their best on summer nights between late June and the end of July.

■ **Habitat** They occur in open and partly wooded habitats, including meadows, heaths, grassy orchards, woodland edges, and parkland.

■ **Search tips** From about 10pm, look out for the lit-up females, which appear as small spots of green light, close to or on the ground. You may find Glow-worms by day, sheltering under rocks or logs.

WATCHING TIPS

Any summertime late-night wildlife watch may produce Glow-worm sightings. Seeing several in one spot creates quite a magical impression, especially if you are out in the quiet countryside, with moths whirring overhead, and all around you the sounds of nocturnal summer wildlife, such as the haunting churr of Nightjars and the croaks and squeaks of Woodcocks roding overhead. Glow-worms are rather scarce and local recorders may be interested in any sightings.

Other beetles

Nettle Weevil

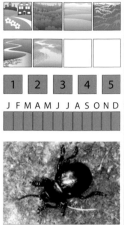

Bloody Nosed Beetle

1	2	3	4	5

J F M A M J J A S O N D

With more than 4,000 species in Britain, beetles offer the sort of overwhelming challenge that some people relish and others simply cannot countenance. Many species are small, nondescript, and barely distinguishable from dozens of others, but there are also some unique and delightful species, both common and rare, that are well worth seeking out. Others are notable simply for being very easy to find and watch. If you prefer a more systematic approach, the UK Beetle Recording website (coleoptera.org.uk) has helpful resources, including a downloadable list of all British species, and a key to all families.

Worldwide, beetles are the most diverse of all animals – one out of every four known animal species is a beetle. The British species represent about 1% of the world total. While many are tiny and very hard to find, we have our fair share of impressive and spectacular species too.

How to find

■ **Timing** Beetles may be seen all year round, though most species are active between late spring and late summer.

■ **Habitat** All types of habitat may be used – a few species are even adapted to live inside houses.

■ **Search tips** You will find more beetles if you use some of the various trapping methods described on pages 234–235 . Otherwise, look for them in foliage, on the ground under stones, in logpiles, under pieces of loose bark on living trees or cut logs, on flowers and in and around water.

WATCHING TIPS

Beetles pursue many different ways of life and are always interesting to watch. Britain has several species of dung beetles, which come to animal droppings, take bits away and store them in an underground nest for their larvae to eat. The click beetles, which may be found on leaves, escape danger not by flying or scuttling but by triggering a clever bodily mechanism that makes a loud click and catapults the beetle into the air. Beetles are certainly fascinating on all levels, and you will easily find numerous species to study, living a stone's throw from your home.

Great Green Bush-cricket

Tettigonia viridissima

Length 7cm

1	2	3	4	5

J F M A M J J A S O N D

This omnivorous bush-cricket is huge when fully grown, and is bright green, with a brown stripe running down its back, the full length of its body. Like other bush-crickets it has very long, fine antennae, and rather small eyes. As with other Orthoptera, nymphs resemble the adult from an early age, gradually growing over a series of moults, but are wingless until their final moult. The adult female has a long, straight ovipositor, which is partly hidden by her long wings (but shows through them). It is quite widespread in southern England and south Wales, but never especially common.

Our largest bush-cricket, this is a truly spectacular insect, and is fairly placid and approachable once you manage to actually find it. Unlike most Orthoptera, it is an omnivore, eating other insects as well as plant material.

How to find

■ **Timing** This species may be seen between May and October, with fully mature individuals appearing in the autumn.
■ **Habitat** It occurs in a range of open, scrubby and wooded habitats, with females in particular often turning up in less typical habitats. It may even visit larger gardens at times.
■ **Search tips** Look for this bush-cricket in trees and fields, and listen for the male's loud stridulations or song, which sound like an old sewing machine in action. Despite its size, it can be hard to spot as it is well-camouflaged and good at keeping very still.

WATCHING TIPS

This is a very impressive insect. The mature male's song is loud and can lead you to his position, but it may still take much searching and staring into the foliage before you actually see him. Mature females may be found wandering away from the bushes and grass as they look for suitable egg-laying sites. If you approach your bush-cricket slowly you should be able to get close for a proper look, but startle it and it will jump clumsily away into cover.

Metrioptera roeselii # Roesel's Bush-cricket

Length 2cm

1	2	3	4	5

J F M A M J J A S O N D

This is an attractive, fairly small bush-cricket. It is mainly dark brown, with orange legs and a green face. The thorax has a cream-coloured edge, and there are two or three yellow spots on the abdomen sides. The female's ovipositor is robust, black-edged, and curved like a scythe. This bush-cricket was confined to the vicinity of saltmarsh habitats on the south-east coast until the early 1900s, when a strong range expansion began. It is now found widely and commonly across south-eastern England and is pushing north and west.

Like other crickets and grasshoppers, this species passes through a series of moults as it grows, and only the final adult stage has wings. The short-winged (brachypterous) form is more common than those with full-size wings.

How to find

■ **Timing** Adults can be seen (and heard) between June and October.
■ **Habitat** This species occurs in varied habitats, preferring sheltered and slightly damp lowlands. You could find it on farmland, meadows with tall grass, the bases of hedgerows, woodland edges and sometimes in parks and gardens.
■ **Search tips** Look for this bush-cricket in grass and other dense vegetation. It is best located by the male's song, a continuous mechanical trilling noise, though is sometimes still difficult to find.

WATCHING TIPS

The non-stop fast trill of the male, recalling the drone of nearby electric pylons, is becoming a characteristic sound of summer in more and more areas. If you live outside south-east England, pass on sightings of Roesel's Bush-cricket to your local insect recorder – photos are always useful to support a record, but in the case of this and other Orthoptera, sound recordings may also help. Most individuals have short wings, but there is a long-winged form that is commoner in some areas than others, for reasons still unknown.

Other bush-crickets, and coneheads

Tettigoniidae

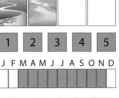

| 1 | 2 | 3 | 4 | 5 |

J F M A M J J A S O N D

Mirollia hexapinna

Long-winged Conehead (juvenile)

The bush-crickets and coneheads can easily be distinguished from grasshoppers by several anatomical features – most strikingly, their very long thin antennae. Also, bush-crickets have spindlier legs, making them good at climbing but less powerful leapers than the short-legged grasshoppers, and female bush-crickets have long ovipositors, which are blade-shaped or curved like scimitars. Of the dozen or so bush-crickets in Britain, some (like Speckled and Oak) are very common and widespread, while rarities include the Sickle-bearing Bush-cricket, which was only found to be breeding in south-east England in 2006.

Less agile than grasshoppers, bush-crickets rely on good camouflage and slow, cautious movements to evade detection by predators. However, when necessary most species can also make a (somewhat clumsy) leap for safety.

How to find

■ **Timing** These insects can be found from mid-spring to late autumn, depending on species, with mature adults around later in the season.

■ **Habitat** Most bush-crickets are at least partly herbivorous and require a reasonable amount of vegetation (for concealment as well as food), so prefer quite lush habitats with long grass and bushes.

■ **Search tips** Listening for the singing males is the best way to locate these insects. They may also be attracted to light, and winged adults often enter buildings in late summer and autumn.

WATCHING TIPS

These slow-moving, well-camouflaged insects are difficult to spot, let alone to watch. Their reliance on camouflage does mean that, once you do find one, it will usually stay put and allow you to look at it closely, or if worried it may start to slowly clamber away rather than leaping. Several bush-cricket species are becoming more common and expanding their range in Britain – pass on any interesting sightings to your local insect recorder, and also consider noting your sightings on iRecord (brc.ac.uk/irecord) - a 'citizen science' resource increasingly used by conservationists and biologists.

Meadow Grasshopper

Length 2cm

1	2	3	4	5

J	F	M	A	M	J	J	A	S	O	N	D

This is probably Britain's commonest and most widespread grasshopper, though not present in Ireland. It is a typical grasshopper in shape, with short antennae and very long powerful hind legs that allow it to quickly and powerfully leap away from any danger, though, unlike other species, it has reduced wings under short wing-cases and is unable to fly. It has a typical long grasshopper head with large eyes set near the top. It is highly variable in colour, with brown, green, violet-brown and intermediate forms, and there is a rare variant that is entirely bright, shocking pink.

It's hard to see how a grasshopper would benefit from being fuchsia-pink (perhaps it could be mistaken for an unusually shaped flower), but the colour variant is frequent enough that it can't have too much of a negative effect on survival chances.

How to find

■ **Timing** You could encounter this species at any point between April and October.

■ **Habitat** It is found in pastures, meadows and other areas of unimproved, damp grassland.

■ **Search tips** The male can be found by his song, a series of short, rolling chirps. A slow and careful exploration of patches of long grass should reveal some Meadow Grasshoppers, though they will bound away from you as you approach unless you move very slowly and carefully.

WATCHING TIPS

Meadow Grasshoppers are quick to leap away if disturbed, so to watch their activity you need to be especially slow and careful in your movements. Sitting quietly in a suitable field should produce good views of them nearby. Noting the numbers you see of the various different colour morphs (and perhaps documenting them with photos – though they are tricky to capture on camera) is an interesting exercise, as this shows marked regional variation.

Other grasshoppers and related species

1	2	3	4	5

J F M A M J J A S O N D

Field Grasshopper

Field Grasshopper

Brown Grasshopper

The Field Grasshopper is about as common as the Meadow Grasshopper, and may occur in the same habitats (but prefers a shorter grass sward). Another nine grasshopper species are present in Britain. Our largest species, the Large Marsh Grasshopper, is only found on peat bogs in southern England. We also have two species of true crickets – the Field Cricket, a large, stout black insect, and the smaller, brown Wood Cricket. Both are scarce and confined to south England. Also on the British list is the bizarre, burrowing Mole Cricket, now extremely rare and possibly extinct. Finally, we have three species of groundhoppers, which resemble tiny, squat-bodied grasshoppers.

Their broadly similar body structure and the tendency for colour variation within species make grasshoppers a difficult group to identify. Often they are more easily recognised by the sounds the males make than appearance.

How to find

■ **Timing** Most species can be seen from spring through to late autumn, as nymphs in spring and adults by mid-summer.

■ **Habitat** Grasshoppers mainly feed on grasses so live in open habitats where at least some grass is present. The groundhoppers are often found near water. Field Crickets occur in open country, while Wood Crickets prefer woodland edges.

■ **Search tips** Grasshoppers and true crickets all produce sounds – a 'song' to advertise their presence and to court mates. Each species has its own particular speed and pitch of chirp, and this helps to locate them as well as to identify them.

WATCHING TIPS

Grasshoppers are highly strung insects and will leap and fly away if you approach them too quickly. If trying to pinpoint a chirping grasshopper, move very slowly and carefully towards its position, stopping and checking from different angles until you spot it. You can download reference sound files for all species from the internet. The website orthoptera.org.uk has links to these, and many other resources for anyone interested in this group of insects.

Snakefly

Length 15cm

1	2	3	4	5

J F M A M J J A S O N D

Snakeflies are peculiar, primitive insects that somewhat resemble lacewings, having large 'tented' wings with a lacy pattern of veins – they are also predatory, like lacewings. The front part of the thorax is elongated, which sets the pointed head well in front of the body and gives the impression of a long, arched, snaky neck. The female also has a long, pointed ovipositor. This species is one of four kinds of snakeflies that occur in Britain. It is found mainly in the south-east, and patchily north and west as far as Yorkshire and east Wales, while other species have a more northerly distribution.

Snakeflies are not often seen and definitely warrant a closer look if you are lucky enough to find one, with their elegant but bizarre body shape. The predatory larvae are also quite distinctive – try looking for them under loose bark.

How to find

■ **Timing** Snakeflies can be seen between April and October.

■ **Habitat** This species is associated with oak trees, so is most likely to be found in and around deciduous woodland. The other species are found in coniferous woodland.

■ **Search tips** Adult snakeflies spend most of their time in the tree canopy. If you check trunks of mature trees you may find females looking for egg-laying sites.

WATCHING TIPS

These strange-looking insects are difficult to find, and consequently are probably under-recorded. Your best chance is finding a female on the tree trunk – she slides her long ovipositor under a flake of loose bark and lays eggs there. The larvae, just as predatory as the adults, live under the bark and hunt small invertebrates for two years or more before they pupate over winter, emerging in spring. Although the four British species belong to four different genera, they are all rather similar. A clear photo of the wings, showing the venation in detail, will help with identification.

Mayflies

Ephemeroptera

Ephemera vulgata

1	2	3	4	5

J F M A M J J A S O N D

Ephemera danica

These famously short-lived insects occur throughout Britain and Ireland, and are represented by about 50 species. All of them are closely associated with water, spending their larval lives there. The adults have long bodies that end in three fine tail filaments. The wings (the hind pair much smaller than the front) are held pressed together above the back, and the mayfly often rests with its rear four legs holding its perch and the front pair stretched straight out in front of it. Adult mayflies do not feed, but form mating swarms over the water.

A substantial mayfly emergence is an impressive sight, and also attracts other wildlife. Insect-eating birds like wagtails and Hobbies and predatory insects like dragonflies may turn up to exploit the bounty.

How to find

■ **Timing** Depending on species, mayflies emerge from the water any time between April and mid-autumn. Because their adult lives are so short, their hatches are usually closely synchronised to maximise the chance of successful mating. So, on one day you may see none and the next day there will be thousands.
■ **Habitat** Mayflies are associated with clean fresh water, including still water but most often rivers and streams. When newly emerged but not ready to mate they often head for nearby fields.
■ **Search tips** Look out for swarms of mayflies dancing over the water on spring and summer days, in their distinctive rising and falling flight style. Also check nearby long grass for resting insects.

WATCHING TIPS

Mayfly nymphs live underwater for a year or more, eating plant matter and sometimes small animals. You will probably catch some when you are pond-dipping. When mature, the nymph crawls from the water and the winged insect within emerges from its skin – but this is not an adult. Mayflies are the only insects that have a winged 'subimago' stage, undergoing one last moult before becoming fully adult. Subimagos are soft-bodied and have cloudy wings. If you find a subimago clinging to a plant stem and fluttering its wings, it may be ready to moult, so keep an eye on it and you could witness a unique transformation.

Lacewings

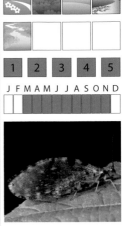

1	2	3	4	5

J F M A M J J A S O N D

These attractive, fragile-looking insects are named for their large wings, with a delicate lattice-work of veins that give them a mesh-like appearance. The body is slender, the legs rather long, and they have small heads with relatively large round eyes and long, fine antennae. Despite their frail appearance, lacewings are avid predators of aphids and other small insects. There are more than 40 species in Britain, of which the most familiar is probably the green, orange-eyed *Chrysoperla carnea* which is very widespread and often comes into buildings in autumn. The Giant Lacewing is our most striking species. It is found mainly in south and west England and Wales, with a few outposts in central Scotland and the Irish coast.

Lacewings are much fiercer than they look and are very much species to be encouraged in the garden, as they will keep down numbers of aphids and other pests. The 'brown' species include the impressive Giant Lacewing, 2.5cm long with very large, dark-spotted wings.

How to find

■ **Timing** Lacewings can be seen from mid-spring through to autumn.

■ **Habitat** They occur in gardens, parks, woodland and wetland habitats. The Giant Lacewing is found mainly in riverside trees.

■ **Search tips** You will usually find lacewings resting within vegetation – long grass, or tree foliage. Also look out for their remarkable eggs, which are laid in small clusters on the undersides of leaves. Each oval egg is at the end of a long, thin stalk.

WATCHING TIPS

Green lacewings hibernate as adults and so are likely visitors to garden bug hotels. The fast-moving, large-jawed larvae, which superficially resemble ladybird larvae are just as predatory and active as the adults. They are stealthy too, often disguising themselves with prey remains to attack clusters of aphids. Adult lacewings are attracted to light and may turn up in your moth trap – they will also enter lit rooms through windows and will hibernate inside buildings.

Pond-dipping

There is no better way to discover freshwater wildlife than pond-dipping. If you have your own pond, then you can hold your own pond-dipping session whenever you like – otherwise, pond-dipping events regularly take place on nature reserves. Even those away from wetland areas often have a dedicated wildlife pond designed for this purpose, although reserves with larger lakes are likely to have more species, while some reserves may offer the chance to try the method in streams, which will hold a different range of species. Pond-dipping is often pitched as an activity for children but really it is for everyone, up to and including professional biologists who use it as a method to survey underwater biodiversity.

Equipment

A good pond-dipping net is heavier and stronger than a butterfly net or sweep net, as it has to cope with possibly getting caught up in water weed or other unseen underwater obstacles. It has a long sturdy handle (telescopic handles are useful) and fine but strong mesh, and most are rectangular and about 20cm across. You will also need shallow white plastic trays in which to place your catch, and clear sample pots or tubes for closer viewing.

Remember to **wash down your wellies** with clean water if you do have to step into the pond.

A gadget called a bathyscope or underwater viewer is a good way of seeing what's moving about under the surface without the problem of surface glare. A more basic bit of kit is a pair of wellies, just in case you need to step into the water to rescue a tangled net. It is also wise to wear old clothes when pond-dipping, and have an old towel handy for your hands.

Technique

Before you start, scoop up some pond water in your trays, ready for when you have some water creatures to add to them. Then it's time to dip. Swirl your net though the water, manoeuvring it gently and slowly through a couple of figure-of-eight loops. Bring it out and turn it inside-out over a tray, so that anything you've caught falls into the water – pick out any stuck bits of weed and put them in the tray too, or back in the pond if too large. Balance the net over the pond just in case any animals are still on it, and then have a look at what you've caught.

Sorting out the pond life

With luck, you will have a number of little creatures moving about in the tray. You can transfer them to a clear tube or a pot with a magnifying lid for a closer look if necessary. You may find tadpoles, or toadpoles or even 'newtpoles', and there is a chance of catching small fish such as sticklebacks, or immature

A shallow white tray is best for checking the water wildlife you've caught.

stages or larger fish (though many wildlife ponds are deliberately kept as fish-free zones as amphibians do much better in small ponds when there are no fish around).

You may catch insect larvae of various kinds. Damselfly nymphs are slim and sinuously wriggly, with three fin-like 'tails', and mayfly nymphs are rather similar but have projections on their sides and longer 'tails' at the tip. Dragonfly nymphs are stout, with short spikes at their tail-tip rather than fin-like structures, and walk rather than wiggle. Diving beetle larvae are long and slim with very well-developed legs and jaws. Some fly larvae, such as those of phantom midges, live in water too – these have no legs and are long, slim and translucent. You could also find caddisfly nymphs, which live inside tubes of gravel and plant matter that they build around themselves. Adult water insects could include diving beetles, water boatmen and backswimmers, and pond skaters, although these are all often too quick for the net. Detailed identification keys are available in field guides and online for the trickier groups, such as the Odonata.

A great range of other invertebrates could turn up too, including the Great Pond Snail and Ramshorn Snail, the former with a pointed shell, the latter a flat curling shell. Water Lice look like slim, long-legged woodlice, while freshwater shrimps are miniature versions of their familiar marine cousins. You could also catch leeches (don't worry, freshwater leeches in Britain are not harmful to humans) and bloodworms, and you may well have tiny organisms such as daphnia and cyclops, though you will need a hand lens or magnifying container to see them properly.

Pond-dipping is often billed as 'for kids', but is **fun and interesting for all ages**.

Tidying up

Place all organisms gently back in the pond when you have finished, and clean and dry all your kit, especially if you intend to use it at other sites. The spread of disease from pond to pond can be a serious problem, so it's important to be scrupulous in this regard. You can pond-dip at your own pond regularly to keep track of seasonal changes, but not too frequently, to avoid excessive disturbance – once a week is about right. Checking out which insect larvae are present will give you advance warning of which species you are likely to see flying near your pond in their adult forms – and if you find newtpoles in the pond then you'll know that there will probably be a newt or two overwintering in some damp shady corner of your garden.

Always **supervise young children closely** around even the smallest pond.

Caddisflies

Adult Common European Caddisfly

1	2	3	4	5

J F M A M J J A S O N D

Caddisfly larva

Caddisflies are related to moths and butterflies, but their opaque wings are hairy rather than scaled. Most are plain brown or grey. The wings are held in a tent shape over the body, and the antennae are usually very long and fine. Caddisflies spend their larval and pupal stages underwater, and most live in a case they build from small bits of underwater debris, bound with silk. There are about 200 species in Britain, and they are hard to identify. One distinct group is the genus *Mystacides*, or 'black dancer' caddisflies – these are shiny black, and males have very large pincer-like mouthparts. They have red eyes, and exceptionally long banded antennae. *Mystacides nigra* is quite common and widespread in Britain.

Caddis larvae use whatever material they find to build their cases – in tanks they can be persuaded to construct pretty shelters from tiny shells and coloured glass. When made of natural materials, the case provides camouflage as well as a protective shelter.

How to find

■ **Timing** Caddisflies appear at various times through spring, summer and autumn, depending on species. Emergences are often synchronous, especially because the majority of species do not feed as adults and must reproduce quickly before their short lives are over.

■ **Habitat** All caddisflies emerge from water and as adults rarely wander far from it. Lakes with lush vegetation at the margins are always good places to look, but some species occur in slow- and fast-flowing water, and some use seasonal pools. These species pass the summer months, when their breeding pools have dried out, in their adult form, spending most of their time resting in treetops.

■ **Search tips** Look out for caddisflies perched in waterside vegetation. They are more active at night, and may be attracted to light, including moth traps – you may also see them by the water in courtship dances late in the day.

WATCHING TIPS

Caddisflies are enigmatic, little-known insects, and not many wildlife-watchers take on the challenge of trying to identify them. Unless you are prepared to kill and dissect specimens for examination under a microscope, the majority that you see will have to go unidentified. Nevertheless, they are rather attractive and easily approached, and there is nothing to stop you taking close-up photos and seeking an expert opinion on what species you might have found.

Stoneflies

Noctuid Moth (*Plecoptera sp*)

1	2	3	4	5

J F M A M J J A S O N D

Yellow Sally Stonefly

These insects are another group considered to be rather primitive – fossils of Plecoptera species from some 300 million years ago are not particularly different to modern stoneflies. They have flattened bodies, with long, strongly veined wings that they carry folded flat over their abdomens. The antennae are long, and there is a fine double 'tail' at the abdomen tip. Stoneflies are strongly tied to clean, fast-flowing water, and so are most likely to be encountered in the uplands of northern and western regions. Britain has 34 species, many of them fairly common, but identification is not easy. Most species are dull brown or grey but there are a couple that are yellow and red.

Stoneflies can be very abundant where conditions are good, and they and their larvae are important food for river- and stream-dwelling fish and birds. They tend to be reluctant fliers but can scamper very quickly over the rocks on their long legs.

How to find

■ **Timing** You might find stoneflies of various species during any month of the year, but most are out in the summer months.

■ **Habitat** These insects are found around the fast-flowing cold, clear streams and rivers where they spend their larval lives.

■ **Search tips** When walking alongside a suitable stream, look out for adult stoneflies resting or basking on rocks close to the water. If you have a careful look under stones at the water's edge, you might find the nymphs, which look like wingless adults.

WATCHING TIPS

The presence of stoneflies indicates very clean water, as the nymphs cannot survive if water quality deteriorates. Their disappearance from streams or rivers is an early warning of pollution. These are rather sluggish insects and weak fliers – it is easy to encourage them onto your hand or into a container for a closer look. The Field Studies Council (field-studies-council.org) publishes a fold-out identification key to the British stoneflies, as well as similar keys to other more difficult invertebrate groups.

Alderflies

Alderfly

1	2	3	4	5

J F M A M J J A S O N D

Alderfly with eggs

Alderflies resemble stoneflies, with their large, prominently dark-veined wings, broad heads, long antennae, flat bodies and sluggish way of moving. However, alderflies hold their wings in a tent shape rather than flattened over their backs. The most common of the species found in Britain is *Sialis lutaria*, which can be found throughout Britain but the other two species that occur here look very similar, and are also widespread. They are primitive members of the insect superorder Endopterygota – those which have a pupal stage rather than moulting straight from larva to adult.

Alderflies have a short flight season and do not feed as adults, instead wandering around on foliage in search of a mate. They provide a valuable food source for nesting waterside birds in spring.

How to find

- **Timing** Adult alderflies emerge in mid-spring and can be seen until mid-summer. You could find the larvae when pond-dipping at other times of the year.
- **Habitat** They occur around still water, including large lakes but also canals and ditches, in both woodland and more open surroundings.
- **Search tips** Look for alderflies in vegetation near water, or warming themselves up on bridge railings.

WATCHING TIPS

Finding the first alderfly of the year is as sure a sign that spring is here as seeing your first Swallow. They are weak fliers and slow walkers, so can be inspected easily. Adults don't live long, quickly seeking a mate, after which the female lays numerous eggs on vegetation at the water's edge. The newly hatched larvae fall or crawl into the water, where they will live for up to two years. You may find them when pond-dipping – they look very like the adults but have spiky bodies, and lack both wings and long antennae.

Ant-lion

Length 3cm

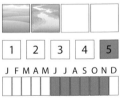

| 1 | 2 | 3 | 4 | 5 |

J F M A M J J A S O N D

Adult Ant-lion

Ant-lion larva

The ant-lions, family Myrmeleontidae, are represented by a couple of thousand species worldwide, but only one is found in Britain and is very rare, confined to small areas of the Suffolk and Norfolk coasts. The adult is a spectacular insect with a long slim black body and long, black-spotted wings with slightly pointed tips – it resembles a damselfly, but has obvious long clubbed antennae and a different head shape. The Ant-lion is best known for the ingenious hunting behaviour of the larvae, which bury themselves in a small pit in sandy earth, and wait for ants and other small creatures to walk in – the loose, steep walls of the pit prevents their escape and the larvae seize them in their fearsome jaws

Adult ant-lions are not as fierce as larvae, eating pollen and nectar plus some live prey. They are more ungainly than the superficially similar damselflies.

Super sites

★ RSPB Minsmere, Suffolk
★ Holkham beach, Norfolk

How to find

■ **Timing** Adult Ant-lions are on the wing between June and September or October. The larval pit traps may be seen outside this time.

■ **Habitat** In Britain, this species only occurs on sandy coastal soils. It is present in the sandy ground by the visitor centre at RSPB Minsmere.

■ **Search tips** The adult Ant-lion is difficult to find as it is only active at night. Look out for the wide conical depressions that mark the trap of a larval Ant-lion.

WATCHING TIPS

If you keep watch over a larval trap until a prey item falls in, you may then see the larva momentarily as it lunges towards its falling victim. Although ants are frequent prey, any small invertebrate that happens along will be attacked – including other Ant-lion larvae, and even sometimes an unfortunate female Ant-lion as she settles on the ground and inserts her ovipositor to lay her eggs. You could find adult Ant-lions in or around pine trees growing at the edge of the dunes.

Water-boatmen, pond skaters and allies

Common Backswimmer

1	2	3	4	5

J F M A M J J A S O N D

Pond Skater

These insects are all in the order Hemiptera or 'true bugs'. They resemble beetles but rather than having a pupa stage they moult from nymph to adult. Aquatic Hemiptera include pond skaters, which have very long legs and walk on water, the surface tension holding their weight; nine species occur in Britain. The Water Measurer, a longer-bodied bug, also walks on the surface, while the spindly Water Stick Insect and squat Water Scorpion hide underwater – both have a breathing tube at their rear ends which they poke out of the water. There are also several species of water-boatmen, which swim just under the surface, mostly on their fronts but the Greater Water-boatman or Backswimmer swims upside-down.

These bugs are all predators with piercing, sucking mouthparts, like a sharp straw. They are quick to attack any non-aquatic insect that falls into the water. Sometimes you will see several Pond Skaters crowded around a drowned large insect, like vultures at a kill.

How to find

■ **Timing** You are most likely to see these insects between April and October but they may be spotted at other times.

■ **Habitat** These insects are found in open, still or slow-flowing water, from small ponds to sizeable lakes, but Pond Skaters especially tend to stay near the shore in sheltered areas. They are strong fliers and quick to colonise new ponds.

■ **Search tips** Pond Skaters are noticeable on calm water, the tips of their splayed legs making dimples on the surface. You'll need to look more closely to see water-boatmen. They tend to swim near the surface but will dive if startled.

WATCHING TIPS

Any of these insects could turn up in the net when you are pond-dipping. Take particular care handling water-boatmen as they can deliver a painful bite with their sharpened-straw mouthparts. Watching water bugs in action in the pond is interesting. Drop a little bit of twig on the water to see how Pond Skaters quickly react to possible prey, and watch Lesser Water-boatmen come to the surface to collect an air bubble, which they use as a sort of aqualung.

Shieldbugs and allies

Common Green Shieldbug

1 2 3 4 5

J F M A M J J A S O N D

Brassica Shieldbug

The shieldbugs are a popular group with wildlife-watchers. They are colourful and comical, easy to watch, they pass through a series of distinct nymph stages and are charmingly gregarious when very small. There are only a few dozen species in Britain, most of which are quite easy to identify. The group includes some rarities, such as the Heath Shieldbug which is found only in southern England, and some recent colonists, including the large and spectacular Western Conifer Seed Bug, a potentially invasive species introduced from America. Some have the shield-shape, but others, as well as most nymphs, are more rounded and beetle-like. Most can be told apart by their colours and patterns.

Most immature shieldbugs have quite a rounded shape and only become shield-shaped in their adult form – others have a round shape even as adults. Many are highly gregarious in their early immature stages, and their mother may also linger to care for them.

How to find

■ **Timing** Most shieldbugs can be seen in immature and then adult stages between mid-spring and mid-autumn.
■ **Habitat** These insects are mostly herbivorous and some are associated with particular plant species, as their names suggest (for example Dock Bug, Juniper Shieldbug and Gorse Shieldbug). In general, they can be found in well-vegetated habitats including meadows, hedgerows, scrub, riversides, marshes, heaths and woodland.
■ **Search tips** Also known as stinkbugs, shieldbugs are generally unappealing to predators and so often wander about on the leaf surfaces in full view, making them easy to find. Look on the undersides of leaves to find clusters of spherical eggs, sometimes guarded by their mother, and small recently hatched nymphs crowded closely together.

WATCHING TIPS

It is easy to get a close look and take photos of a shieldbug for later identification. The nymphs are also frequently encountered, so you can soon build up a library of images of the commoner species. Shieldbugs are among the very few insects to offer any kind of parental care for their young, and this is fascinating to observe. One species is even known as the Parent Bug because of the devotion of the mothers, who guard the eggs and then supervise the cluster of newly hatched young while they feed, staying in attendance until they have passed through their third moult.

Other true bugs

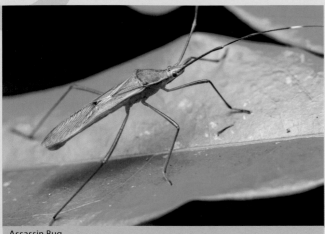

Assassin Bug

| 1 | 2 | 3 | 4 | 5 |

J F M A M J J A S O N D

Froghopper

There are about 1,700 species of true bugs in Britain overall, and they are an exceptionally varied bunch. We have already seen some of the aquatic, predatory species, and the sap-sucking shieldbugs. Others include the froghoppers, responsible for the blobs of frothy 'cuckoo-spit' you'll see on low plants through spring, within which the nymphs shelter. There are the slim-bodied, predatory assassin bugs and damselbugs, and delicate, plant-eating capsid bugs, and the aphids and whiteflies that can be such a problem in the garden – but attract interesting predators like lacewings and ladybirds. You can find examples of all with a little careful searching.

The true bugs are some way behind the beetles in terms of species diversity, but show just as much, if not more, variety in appearance and lifestyle. While many look rather beetle-like, others have obvious exposed wings and could be mistaken for flies.

How to find

■ **Timing** As with most other insects, the period between May and September sees peak diversity but you might find a few species at other times of year.
■ **Habitat** True bugs live in all kinds of habitats. Anywhere where the vegetation is dense and varied will hold a good range of species.
■ **Search tips** Look for these insects on flower-heads and within foliage.

WATCHING TIPS

The true bugs are fairly typical of insect orders in Britain – there are a few easily recognised and common species, and a great many more that will take time and skill to find and identify. 'Beating' vegetation (see p234) is a good way to find bugs that are on the shy side, though be aware that many species are very fleet-footed or excellent leapers, and will rapidly escape from your collecting sheet. Sweep-netting will also produce various bug species. There are many hemiptera enthusiasts in Britain – have a look at britishbugs. org.uk for some useful resources.

Cicadetta montana

New Forest Cicada

Length 2.5cm

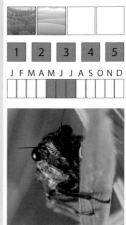

| 1 | 2 | 3 | 4 | 5 |

J F M A M J J A S O N D

Many of us are familiar with cicadas of various kinds from holidays in hotter climates. Just one species has ever been found in Britain and it is extremely rare. In appearance it is a typical cicada – a large, squat and fat-bodied insect with a blunt head, short antennae, bulbous dark eyes, and very large, strongly veined wings that reach well beyond its abdomen tip. The abdomen has narrow orange bands and the wing edges and legs are also orange. This cicada is only found in the New Forest, and its toehold here is precarious. Wildlife-watchers are encouraged to help scientists map and monitor its distribution, to help ensure its continued survival.

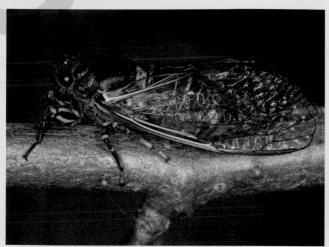

Its size, rarity and uniqueness make this insect top of many wildlife-watchers' wish-lists, and its New Forest habitat hosts an array of other interesting wildlife too.

Super sites

★ New Forest, Hampshire

How to find

■ **Timing** Adult New Forest Cicadas are only around for a few weeks each year. You can see them between late May and early July, on warm sunny days when the air temperature is at least 20°C. The nymphs live underground for up to eight years.

■ **Habitat** It is found in open sunny and sheltered places near trees, including glades, wide woodland paths, and scrubby woodland edges, ideally south-facing.

■ **Search tips** The New Forest Cicada can be located by the male's song, a continuous ringing trill so rapid it sounds like a hiss, which is unfortunately too high-pitched to be audible to some older people. It can, however, be picked up by a bat detector. The insects sing from a perch in a low bush, and can be very difficult to find as they hide well and sit very still.

WATCHING TIPS

Finding the cicada is not easy, but worth it to see this most rare and spectacular of British insects. Your observations can also help to improve knowledge about the species. Conservationists working to protect it have launched a smartphone app that uses the phone's microphone to pick up the sound – you can download the app, and find out much more about the New Forest Cicada and the ongoing work to help save it from extinction, at the project's website: newforestcicada.info.

Wasp Spider

Argiope bruennichi

Female Wasp Spider

Length 18mm

1	2	3	4	5

J	F	M	A	M	J	J	A	S	O	N	D

Web stabilimentum

The Wasp Spider is our most striking orb-weaver spider, both large and colourful – and it also builds a particularly unusual web. A fully-grown female has a leg-span easily reaching 3cm and shows bold black, white and yellow stripes on her oval abdomen, with yellow, black-banded legs. The male is much smaller and less striking and is usually consumed by the female after mating. This spider is native to the Mediterranean but has been present in Britain for some years, slowly spreading north and west from the south-eastern counties.

Wasp Spider egg-sacs are large (2.5cm across), pale brown and flask-shaped; you could find them in the vegetation, stuck in place with silk.

Super sites

★ RSPB Rainham Marshes, Essex
★ RSPB Lakenheath Fen, Norfolk
★ RSPB Arne, Dorset

How to find

■ **Timing** Well-grown Wasp Spiders can be seen from early summer, but it is not until early autumn that the females have reached their full magnificent size. By mid-October most have died, leaving behind an egg-sac, from which new spiderlings emerge next spring.

■ **Habitat** Wasp Spiders are usually found in long grass in sheltered places. The web is often built low to the ground, within the grass sward.

■ **Search tips** It is easiest to find in September when the females are at their biggest. Look low into the grass at any spots where you might imagine a person has stepped, leaving a small 'hollow' of squashed grass. The female will usually be in the centre of her web, but an unoccupied web can easily be recognised by its stabilimentum, an obvious zig-zag structure of thick silk running down the centre.

WATCHING TIPS

These beautiful spiders are a fascinating recent addition to our fauna. If you find a female earlier in the season, you may be able to discover a male living nearby, waiting for the female to complete her final moult. This is the time when he stands the best chance of mating without being eaten. He may also leave part of his pedipalps (the organ used to transfer sperm) stuck in the female's reproductive tract after mating – this helps prevent other males from mating with her, though does not improve his chances of escaping with his life after the act.

Dolomedes plantarius, D. fimbriatus

Fen Raft and Raft Spiders

Raft Spider

Length 2.3cm (Fen Raft)
2cm (Raft)

1	2	3	**4**	**5**

J	F	M	A	M	J	J	A	S	O	N	D

Fen Raft Spider

These two uncommon and striking spider species live by water and hunt by using the water as a sort of web, feeling for surface vibrations that indicate prey. The Fen Raft Spider is very rare, found only at specific sites in East Sussex, East Anglia and Wales. Reintroductions are helping to increase its range. The Raft Spider is more widespread with some sites in northern England and Scotland, but still very localised and uncommon. Both spiders have dark, streamlined bodies with a cream edge, and long legs which they spread wide when they walk on the water's surface. A similar-looking species that walks on water is the Pirate Otter Spider, a type of wolf spider. It is much smaller (9mm).

Super sites

★ Redgrave and Lopham Fen, Suffolk (Fen Raft)
★ Castle Marshes, Suffolk (Fen Raft)
★ Pevensey Marshes, East Sussex (Fen Raft)
★ Crymlyn Bog, Swansea (Fen Raft)
★ Thursley Common, Surrey (Raft)

How to find

■ **Timing** Fen Raft Spiders may be seen from April through to October, Raft Spiders from May to August.

■ **Habitat** The Raft Spider is mainly found around pools with sphagnum moss in boggy heaths and woodland. The Fen Raft Spider is found in more open marshland, around ditches and small ponds.

■ **Search tips** Look for the spiders on open water – it is difficult to spot them when they are waiting at the water's edge but when they go out onto the surface they are easy to see.

WATCHING TIPS

These spiders do not make webs to catch prey but hunt on foot. A hunting raft spider sits on floating vegetation by the water, with its front appendages in contact with the surface. When it senses vibration it can quickly run out to investigate – its well-spread legs mean that the surface tension of the water can support its weight. If it finds prey it delivers its venomous bite. Both species can tackle large prey, even tadpoles and small fish, and can swim or dive underwater if they need to. The Fen Raft Spider is protected by law because of its rarity.

Crab spiders

Misumena vatia

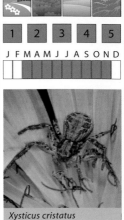

| 1 | 2 | 3 | 4 | 5 |

J F M A M J J A S O N D

Xysticus cristatus

The name 'crab spider' describes a variety of species which are more or less crab-like in shape, and use camouflage to disguise themselves in open situations, striking when prey moves close enough. The species most often referred to by the name 'crab spider' is *Misumena vatia*, a round-bodied spider can change its colour between white, pink and yellow, to match the flowers on which it sits. Other common British species include Common Crab Spider, found commonly throughout Britain, and Green Crab Spider, which is mainly found in southern and central England. A couple of dozen other species occur in Britain, some very rare (for example, *Ozyptila pullata* is known from just a single chalk quarry in Kent).

Crab spiders are the ultimate ambush hunters, waiting motionless with arms spread, for a fly or bee to blunder into their embrace. Check Ox-eye Daisies in late spring for a good chance of finding the common *Misumena vatia*.

How to find

■ **Timing** These spiders may be seen from mid-spring through to early autumn, with May, June and July the peak months.
■ **Habitat** Crab spiders occur in all kinds of lushly vegetated habitats that support plenty of insects.
■ **Search tips** Look for *Misumena vatia* on open-structured flowers, such as Ox-eye Daisies or Cow Parsley. Other crab spiders may use flowers or foliage to hide in. You may spot a crab spider when you notice a fly or other insect on a plant or flower that is not moving and is in a curiously unnatural position – a closer look then reveals the camouflaged crab spider holding onto it with its front legs. Crab spiders may be caught by beating, or using a sweep net (see p234–235).

WATCHING TIPS

The camouflage of *Misumena vatia* is most impressive. Most individuals seen are white, but yellow individuals are often seen too – search yellow flowers to find them in their yellow phase. The spiders can change colour when they move to different-coloured flowers but the process takes a long time – going from white to yellow can take 20 or more days, although the transformation is much quicker in reverse. The largest, roundest individuals are female; males are much smaller and very active in their search for females.

Argyroneta aquatica

Water Spider

Length 15mm

1	2	3	4	5

J F M A M J J A S O N D

When seen on land, the spider's abdomen has a velvety appearance because of the thick hair, which is used to trap air to carry to the 'diving bell'. Studies show that the underwater air bubble in the bell draws in extra oxygen from the water.

This medium-sized species, also known as the Diving-bell Spider, looks fairly unremarkable at first glance – a quite robust dark-bodied spider with strong, long legs. However, its underwater lifestyle sets it apart from other spiders, and it is the only British representative of its family. The Water Spider has an extensive but patchy British distribution – it is most abundant in the south-east and north-east of England but occurs into the far South-West, west Wales and north Scotland, as well as Ireland.

How to find

■ **Timing** This spider can be seen mainly between April and October.

■ **Habitat** It is found in lakes, ponds, and sometimes very slow-flowing rivers and streams.

■ **Search tips** This spider lives almost full-time underwater. It is, therefore, hard to see in the wild. You may catch one while pond-dipping – it will appear silvery, and will quickly dive underwater when you place it in a water-filled collecting tray. Using an underwater viewer (see p250) may allow you to spot one, in its 'diving bell' within clumps of underwater vegetation.

WATCHING TIPS

The Water Spider is difficult to watch in wild conditions. If it is present in your garden pond, you may be able to keep one in an aquarium filled with pond water for a short time, to watch its remarkable behaviour. Like other spiders, it breathes air, and creates its 'diving bell' by making a tight silk mesh bound to underwater plants. It then fills this with air, using the bubbles that become trapped in its body hairs when it surfaces. It lives and even has its young inside its diving bell, and feeds on underwater prey that it rushes out to catch when it senses them touching the sides of the bell or its anchor threads.

False Widow Spider

Steatoda nobilis

S. nobilis

Length 14mm

1	2	3	4	5

J F M A M J J A S O N D

S. bipunctata

This spider, more properly known as Noble False Widow, is native to Madeira and the Canaries, but has spread through western Europe, and has been present in Britain since before 1900. It is found widely across southern England and is spreading north, with records across East Anglia, the Midlands and south Wales, and a handful from further north. It has long legs and a very round, glossy black body with pale markings. There are several other *Steatoda* species in Britain, of similar appearance, including the very widespread, red-bodied *S. bipunctata*, the rare and localised *S. albomaculata*, and the fairly common *S. grossa*, which is frequently confused with *S. nobilis* and does come into buildings.

The *Steatoda* spiders have round bodies with a polished appearance. The small males are short-lived but females can live for five or six years. They can produce several egg sacs each year if there is a good food supply.

How to find

- **Timing** You may find this spider at any time of year.
- **Habitat** This species and several of its relatives are most often found on buildings, though they rarely come indoors.
- **Search tips** Look out for the small, irregular tangle web, built close to some kind of shelter, such a crack or gap behind a pipe. This is where the spider builds its shelter, a silk tunnel from where it can monitor its web. You are more likely to see the spider out and about in the evening and night-time.

WATCHING TIPS

This species has attracted much adverse publicity in recent years, but reports of the deadliness of its bite are greatly exaggerated. All spiders have a venomous bite, and the false widows are among the few who can actually penetrate human skin and deliver a painful bite, not unlike a wasp sting – but suffering a bad reaction is very rare. Being bitten in the first place is also rare, as these are shy and docile spiders, and will do no harm at all if left alone. The females in particular are attractive and long-lived arachnids.

Zebra Spider and other jumping spiders

Length 7mm

| 1 | 2 | 3 | 4 | 5 |

J F M A M J J A S O N D

Zebra spider

Heliophanus auratus

The jumping spiders are extremely appealing, especially when viewed front-on when their very large front pair of eyes are visible. The Zebra Spider is the most familiar species, being found on buildings and having a distinctive black-and-white striped appearance (though individually the pattern is quite variable). It does also occur in wilder habitats. Some 40 other species of jumping spiders are known from the British Isles. Scarcer species include *Heliophanus auratus*, known only from a few shingle beaches in Dorset and Essex – RSPB Arne is a site for this rarity.

With their ability to return our gaze with an appealingly wide pair of eyes (though like other spiders, they do have extra eyes besides the main pair), jumping spiders have a good deal of charm and can even win over committed arachnophobes.

How to find

■ **Timing** Zebra Spiders can be seen from March to October. They are diurnal and hunt particularly actively on sunny days.

■ **Habitat** Look for Zebra Spiders on walls and fences. Other jumping spiders can be found in all kinds of habitats.

■ **Search tips** These spiders hunt by sight, so are most likely to be spotted in open areas that give them a good field of view, such as on rocks or tree trunks. Check sunlit walls and wooden fences on summer afternoons and you stand a good chance of spotting a Zebra Spider. Their colours can provide good camouflage against wood and mortar but you'll notice them when they move in their rapid stop-start manner.

WATCHING TIPS

Jumping spiders hunt their prey by stalking and then, as the name suggests, jumping on it. Because the prey is mostly flying insects, these spiders need to be extremely quick and they are – they will also quickly jump away if alarmed. However, if you give them space and don't startle them, you can easily watch them in action and take photographs – though they may spot themselves reflected in the lens, and attack the camera! Although jumping spiders don't make webs, they do use a silk safety line to anchor themselves to the wall, so if the jump fails they can climb back up.

Other spiders

Common House Spider

1	2	3	4	5

J F M A M J J A S O N D

Woodlouse Spider

With more than 650 species recorded from the British Isles, as well as some 30 harvestmen, this group of animals rewards patient study. Familiar species include the Garden Spider, an orb-weaver common in gardens, which grows to impressive sizes by autumn, and the Common House Spider, which lives almost exclusively in buildings and famously dashes across our living rooms. Look a little further to find Four-spotted Orb-weavers, whose fat bodies make them our heaviest spider; wolf spiders, hunting on foot through the leaf-litter; Long-jawed Spiders which hide under leaves, their long legs stretched out in front and behind; and Tube-web Spiders, hiding in silk-lined tunnels with iridescent green fangs.

We don't think of spiders as very colourful animals, but the Woodlouse Spider is vivid red-pink. Another colourful species is the aptly named Ladybird Spider, a burrowing species which occurs only on Dorset heathland. Reintroductions are gradually increasing its numbers.

How to find

- **Timing** Most species are easiest to find in summer and autumn, but many are active all year.
- **Habitat** Spiders occur in all kinds of habitats. Anywhere that has plenty of insects will always be particularly good for them.
- **Search tips** Orb-web spiders are usually spotted quite easily, at least when guarding the centres of their webs. Others may be found by slow, careful searching of vegetation and turning over rocks and logs. You can also catch them in pitfall traps or through beating and sweep-netting. A few species can bite so always handle with care.

WATCHING TIPS

For every committed arachnophobe, there is a spider enthusiast, and there are many resources to help you get to find and know our species, from Facebook groups to the British Arachnological Society. Start by seeing which species you can find in your garden. Spiders show complex behaviours, from the tricks males use to avoid being devoured by their mates, to their many hunting methods, the wonders of web construction and the aerial dispersal of spiderlings, floating on parachutes of silk. This is an incredibly rewarding group to study, and even those nervous of spiders will find their fears dissipate as their understanding grows.

Austropotamobius pallipes

White-clawed Crayfish

Length 12cm

| 1 | 2 | 3 | 4 | 5 |

J F M A M J J A S O N D

This is Britain's only native crayfish, but is under threat because of a larger, invasive species – the Signal Crayfish, introduced from North America. Signal Crayfish outcompete and prey on the native species, and pass on disease, in particular the deadly 'crayfish plague'. Two other non-native crayfish species are also now established in Britain. Our native crayfish is a grey-green, lobster-like crustacean with large claws and a lobed tail-end, and the undersides of its claws are white. The Signal Crayfish is larger and redder – the undersides of its claws are bright red. The White-clawed Crayfish is still quite widespread in central and northern parts of Britain, and in Ireland where the Signal Crayfish is not established.

Checking claw colour is the best way to be sure you are looking at our native crayfish and not an introduced exotic species. Very big crayfish are almost certainly going to be Signals, as the native species doesn't grow to a great size.

How to find

■ **Timing** This crayfish is most likely to be seen in warmer months but could be encountered at any time of year – courtship and mating occurs in late autumn, and females carry their eggs through the winter. It is nocturnal.

■ **Habitat** White-clawed Crayfish are found in clean freshwater habitats, from small streams to reservoirs, with rocks and other hiding places.

■ **Search tips** Look out for these crayfish foraging in the shallows, especially around dusk and dawn. They are also sometimes accidentally caught by anglers using sinking baits.

WATCHING TIPS

The White-clawed Crayfish has been designated as Endangered on a global scale, with introduced crayfish species threatening it across its range in Europe as well as in Britain. Therefore, keeping a very close eye on its fortunes – and the spread of Signal and other non-native crayfish – in Britain and especially Ireland is very important. Many crayfish surveys and conservation projects are underway. You can find out more on Buglife's website: buglife.org.uk/crayfish-projects-near-you.

Hermit Crab

Pagurus bernhardus

Length 3.5cm

| 1 | 2 | 3 | 4 | 5 |

J F M A M J J A S O N D

This fascinating crab can be found around all British and Irish shorelines. It is instantly recognisable in that it occupies an old shell of a gastropod mollusc, such as a periwinkle or whelk. When you see a shell like this moving along at (relatively) high speed, you can be reasonably certain you have found a Hermit Crab, and a close look will reveal claws, legs and beady eyes protruding from the shell's opening. Unlike most crabs, its body is soft and flexible, allowing it to squeeze into its makeshift home.

Hermit Crabs seek out shells that are big enough to tuck their whole bodies inside for protection, but their continual growth forces them to upgrade their home periodically. Sometimes two will both want the same shell, and a pushing and shoving battle ensues.

How to find

■ **Timing** Look for Hermit Crabs around low tide.

■ **Habitat** You are most likely to find Hermit Crabs on rocky shores with rockpools, especially near the low tide mark but quite often higher up as well.

■ **Search tips** Rockpooling should soon produce a Hermit Crab or two. Also, if you see a gastropod shell resting in an unexpected place, pick it up and look inside – you may then see the tucked-up legs of a Hermit Crab deep inside. Left alone for a little while, the shell will soon sprout legs and scuttle away.

WATCHING TIPS

These delightful little animals look comical as they scuttle along with their bulky shell bobbing along behind their fast-moving legs. If you are very lucky you could see one changing shells, a necessary process as its body grows. It will find a suitable shell first, explore it carefully, and then (as long as the shell is suitable), quickly draw itself out of the old shell and then manoeuvre itself backwards into the new one, giving you a glimpse of its long, soft body that wiggles around the internal coil of the shell.

Carcinus maenas

Shore Crab

Width 8cm

1	2	3	4	5

J F M A M J J A S O N D

This is our most common crab species, and can be found in abundance around all coasts throughout Britain and Ireland, including estuaries, and sometimes around coastal lagoons. It is a relatively small crab even when fully grown, with the familiar shield-shaped carapace that is wider than it is long, and fairly large claws. Adults are usually a dull brown-green or reddish-brown colour, but the tiny juveniles are more varied, with brighter green, white and marbled forms.

If you pick up a crab, hold its carapace on both sides, gripping the body behind the front legs to keep your fingers out of reach of those pinching claws. Even small crabs are capable of giving a painful nip.

How to find

■ **Timing** You can find these crabs at all times of year. Breeding occurs at any time of year but peaks in summer. They are most active at night.

■ **Habitat** These crabs occur on all kinds of coastlines. Juveniles are common in rockpools, while adult females seek sandy ground in which to place their eggs.

■ **Search tips** You are most likely to find small juvenile Shore Crabs while rockpooling, but look out for larger individuals on open beaches, perhaps scavenging in the strandline. You can also catch them at high tide by 'crabbing' – attaching some smelly bait such as bacon to a long line, dangling it in the water and waiting. Scoop up your catch in a net, and transfer to a bucket of seawater if you wish, for a closer look. Take care if handling them, as they can and will nip.

WATCHING TIPS

If you carefully check underneath stones in large, weedy rockpools you should find reasonable numbers of baby Shore Crabs, exhibiting a range of colours and patterns. When released, they quickly squeeze back into a hiding place among seaweed or back under another rock. Shore Crabs are predators and scavengers, and have extremely broad diets. In other parts of the world, where they are introduced and invasive, they have harmed populations of animals as diverse as clams, fish and even other crabs.

Other crustaceans

Water Louse

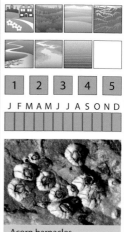

1	2	3	4	5

J F M A M J J A S O N D

Acorn barnacles

When we think of crustaceans, we tend to picture lobsters and crabs, but the group also includes barnacles and woodlice, planktonic animals like water fleas and copepods, and a great variety of shrimps that live in both freshwater and the sea. Woodlice are the most familiar land crustaceans, but they cannot survive away from wet environments, and they have fully water-dwelling cousins such as the Water Louse and the Sea Slater. A variety of crab species can also be found around our coasts, as well as Common and Squat Lobsters. You should find a variety of freshwater and marine crustaceans through pond-dipping and rockpooling respectively.

Only when underwater do Acorn Barnacles open up their protective plates and reveal their feeding appendages. When the tide is out, the plates protect against dessication. Their name comes from the outer plates which resemble an acorn's cup – several species occur around the British coastline.

How to find

■ **Timing** These animals may be found at any time of year.
■ **Habitat** Depending on the species you're after, you'll need to look on shorelines and in rockpools, in ponds and streams, and within rotting woodpiles.
■ **Search tips** The appropriate search methods for these animals vary according to species. Many crabs are shy and nocturnal, and quick to scuttle into hiding places if disturbed. Microscopic crustaceans can be found by pond-dipping, and inspecting the water with a strong hand lens.

WATCHING TIPS

Crustaceans are, by and large, difficult to watch. Many are very small indeed, and many of the larger species are shy, spending most of their time buried in sand or out at sea, or are aggressive, discouraging close approach and handling. Nevertheless, a few hours at a pond or seaside, equipped with plenty of containers of varied sizes, should allow you to see and watch some of them. The tiniest species, such as copepods and water fleas, are often exquisitely beautiful in an otherworldly way – use clear pots and a hand lens to watch the various curious ways they swim and to study their peculiar anatomy.

Anemones

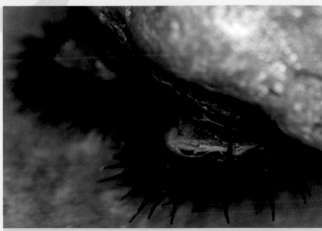

Beadlet Anemones

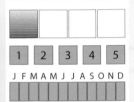

| 1 | 2 | 3 | 4 | 5 |

J F M A M J J A S O N D

Snakelocks Anemone

Sea anemones are mostly sessile (fixed to one place) relatives of the jellyfish. When submerged and feeding, they show a mass of soft tentacles, which wave gently in the water and ensnare passing prey, subduing it with stings before drawing it into the animal's body. When the tentacles are drawn in, the anemone becomes just a jelly-like blob. Several species can be found in the intertidal zone and they are one of the delights of a good rockpool. The small, red Beadlet Anemone is one of the most common, while the Snakelocks Anemone with its flowing green red-tipped tentacles is probably the most beautiful.

Anemones are beautiful and quite alien with their crown of softly waving tentacles. Size and colour is key to identifying the 70 or so species found in British and Irish waters. The green tint of most Snakelocks Anemones comes from a photosynthesising alga that lives in its body tissues.

How to find

■ **Timing** Search deep rockpools at low tide for the best chance of finding plenty of anemones.

■ **Habitat** These organisms are not confined to rockpools but this is the easiest place to find them. However, if you snorkel in sheltered open bays you will probably find more and larger ones.

■ **Search tips** The waving tentacles are usually easy to spot. Some anemones can tolerate being exposed to the air for short spells by pulling in their tentacles – these may be spotted on rocks close to the low-water mark.

WATCHING TIPS

Sea anemones are predatory, the undersea equivalent of carnivorous plants, waiting for prey to blunder into their flower-like tentacles. Watch for long enough and you could see this happen, if a shrimp or small fish strays too close. The tentacles fire stinging cells at the victim, delivering paralysing neurotoxins that quickly immobilise the prey. Then tentacles plus prey are slowly drawn into the anemone's body and a lengthy digestion process begins. Never attempt to prise sea anemones from their bedding places – there is no need and you may damage them.

Leopard Slug and other slugs

Limax maximus

Length 20cm

1	2	3	4	5

J F M A M J J A S O N D

Leopard Slug

Arion rufus

The Leopard Slug is our largest species of slug. It is also known as the Great Grey Slug, and the name that best suits it depends on the individual – some are yellow with distinct dark spots, while others are much duller. Like most slugs, it has a saddle near the front of its body, two pairs of tentacles on its head, and an obvious breathing hole close to its saddle. This slug is common in Britain, and is one of about 30 species present in the British Isles. Unlike some of the others, it is not a pest in the garden, mainly feeding on algae growing on rotting wood.

Slugs may not be very charismatic at first glance, but they fill important roles in ecosystems and have their own range of interesting (if slow) behaviours. Their presence also attracts predators, such as shrews and Hedgehogs.

How to find

■ **Timing** Leopard Slugs and other slugs can be found all year round.
■ **Habitat** The Leopard Slug is most often encountered in woodland and similar habitats. Several slug species can be found in gardens, where they provide food for Hedgehogs, thrushes and many other animals.
■ **Search tips** Slugs need to remain damp so keep close to leaf litter, or stay under rocks or logs. They are more likely to appear at night, and after rainfall.

WATCHING TIPS

The mating of the Leopard Slug is one of the real marvels of the British countryside. You are most likely to see it in summer, in damp, shady woodlands. As with snails, they are hermaphrodite, so both individuals play both roles. After prolonged courtship, the pair move together to a raised spot, and then lower themselves down on a thick thread of mucous. Entwined together, they evert their reproductive organs, which are large, frilly, translucent blue and bizarrely beautiful, and through contact between these, exchange sperm. Then they separate and climb back up their mucous rope, to later lay several hundred pearl-like eggs in damp soil.

Roman Snail

Width 5cm

| 1 | 2 | 3 | 4 | 5 |

J F M A M J J A S O N D

Garden Snail

This is a very large and rare snail, which may have been introduced to Britain in Roman times, as a food source. Other snails are just as edible, but this species' size made it a particularly popular choice – it is, however, now legally protected in Britain. Many other land-dwelling snail species occur in the British Isles, including the familiar Garden Snail, and the common and attractive banded snails. Rare species found in Britain include the Plaited Door Snail, a tiny woodland species with a long corkscrew shell, which made the news in 2011 after being rediscovered in Scotland more than 100 years after it was presumed to have gone extinct.

Because they can protect themselves in dry conditions by withdrawing into their shells, snails can be found in the open whatever the weather. You'll often find clusters of banded snails on the stalks of tall plants like hogweeds.

How to find

■ **Timing** You could find Roman Snails and others at any time of year. They are most active at night and after rainfall.
■ **Habitat** The Roman Snail is found on chalk downland in southern Britain. Other snails occur in a wide variety of habitats, their shells enabling them to endure drier conditions than slugs can manage.
■ **Search tips** Walking on downland, you'll quite often find the empty and bleached shells of long-dead Roman Snails. Locating living animals is more difficult – look in undergrowth and other sheltered damp spots, or visit sites just after heavy rain.

WATCHING TIPS

If you are willing to sit out on a downland slope in the rain on a springtime night, possibly for several hours, you could see Roman Snails performing their surprisingly prolonged and tender courtship, with much gentle touching of tentacles before they press themselves together along their full underside length to copulate. The snail fauna of Britain in general is not especially well known, and there is plenty of scope for amateurs to make useful contributions to knowledge of distribution and behaviour.

Rockpool gastropods and bivalves

Flat Periwinkle

Common Mussel bed

1 2 3 4 5

J F M A M J J A S O N D

A variety of gastropods and bivalves may be found in and around rockpools around our coasts, some of them quite exquisitely beautiful. Because these animals have a protective shell to keep them from drying out, they don't need to be immersed in water full-time and may be found on rocks that are exposed to the air at low tide. Common species you'll find on most coasts include the colourful Flat Periwinkle, several species of limpet, Common Mussels and the Netted Dog Whelk with its pointed, bumpy shell. Among the more localised species are the colourful Painted Top Shell and Flat Top Shell, and the delicate, elongated Needle Whelk.

Many of the more unusual species of rockpool molluscs are found only on western or south-western coasts. Among the more distinctive species is Variegated Scallop, which has white, pink, red, orange, yellow and purple forms, and may be seen swimming in large rockpools.

How to find

- **Timing** Visit the coast at the lowest tide possible for the chance of seeing a wide range of species.
- **Habitat** The best places to look are shores with a wide range of habitat types, including some sandy patches, steeper and more gently sloping rocks, and rockpools of varied depths.
- **Search tips** Look carefully on all exposed parts of rock right down to the waterline, using a torch to look into crevices. Don't pick shells from the rocks, but take photographs for later reference.

WATCHING TIPS

These animals are mostly slow-moving, if they move at all, but the real joy of them is in appreciating the interesting structure of their shells and their often quite beautiful colours and patterns. However, a few species are quite mobile too. Watch out for winkles gliding sedately across the rock, grazing on algae, and scallops swimming with surprising speed and grace, by opening and closing their shells to push out water and create propulsion.

Other marine molluscs

Grey Sea Slug *Aeolidia papillosa*

Sea Hare *Aplysia punctata*

Not all molluscs wear their shells on the outside. The group also includes soft-bodied, highly mobile animals that have an internal shell – the cephalopods (octopuses, squid and cuttlefish). The Sea Hare and sea slugs (nudibranchs) are also molluscs – in fact they are technically gastropods even though they have no shells in their adult form. These animals need to stay underwater to avoid desiccation, and most are found in the open sea, but some may turn up in rockpools. Some of the sea hares (nudibranchs) in particular are among the most spectacular rockpool animals of all.

Nudibranchs come in a dazzling array of colours and patterns, and some have extravagant tentacles and other body ornamentations as well, lending a tropical feel to large rockpools on the west coast of Britain.

How to find

■ **Timing** Visit as the tide is on its way out to give yourself the most time to look.

■ **Habitat** The deepest rockpools, closest to the low-tide mark, are most likely to hold interesting species.

■ **Search tips** Look carefully into the water (try an underwater viewer) and gently move items of substrate, replacing them afterwards in their exact original position. Look out for any sudden movements. Bear in mind that some cephalopods and nudibranchs are masters of disguise, with some even able to change their colour to blend in perfectly with their surroundings.

WATCHING TIPS

Cephalopods are rare in rockpools but you could also see them while snorkelling. They are fascinating, intelligent animals – expert predators with amazing hiding skills. The sea slugs (nudibranchs) around our coasts include some dazzling species. The colours may be a warning – for example, the stunning Scarlet Lady, a nudibranch covered with red tentacles, can store stinging cells from its jellyfish prey and use them against its own enemies.

Jellyfish

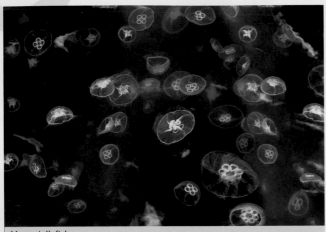

Moon Jellyfish

| 1 | 2 | 3 | 4 | 5 |

J F M A M J J A S O N D

Lion's Mane

Jellyfish are not generally seen unless you are in or on the sea, but they often come to the surface and so can be seen easily from wildlife-watching boat trips, offshore and in harbours and sea-lochs. Most of them possess tentacles that can deliver painful stings, so direct contact is best avoided. Species seen around Britain include the translucent Moon Jellyfish, with its reproductive organs prominently visible as four ring shapes inside its bell, and the Lion's Mane Jellyfish, a large pink species with a thick, trailing mass of pink tentacles. The Compass Jellyfish, with distinctive black radiating stripes on its yellow bell, is another common species found mainly off south and western coasts.

The large, spectacular Lion's Mane jellyfish has a very extensive world range, occurring right across the northern Atlantic and Pacific Oceans, and possibly also in the southern Hemisphere. Its bell can reach up to 2 metres in diameter.

How to find

- **Timing** Jellyfish are most abundant in our waters in the warm, summer months.
- **Habitat** These animals swim at all depths, but you are most likely to see them near the surface in sheltered, shallow seas.
- **Search tips** If you take a rowing boat out into a sheltered inlet, look out for jellyfish drifting past you. And beware of them when swimming, particularly in sea lochs and other sheltered waters.

WATCHING TIPS

Jellyfish have a bizarre, otherworldly beauty to them, especially the Moon Jellyfish with its transparent body that reveals that the way it is built is nothing like that of animals we know well. Their structure is actually very simple, with no true nervous or digestive system, and they catch their prey passively. Their movement through the water, though, is hypnotically graceful. As long as you avoid touching them, they are fascinating to watch at close range.

Sea stars, sea urchins and allies

Common Starfish

Edible Common Sea Urchin

1	2	3	4	5

J F M A M J J A S O N D

The phylum Echinodermata includes sea stars (also known as starfish), sea urchins, crinoids, brittlestars and sea cucumbers. They are simple animals with spiny or spiky skins, and many look more like flowers than animals, with a central body surrounded by symmetrical radiating 'arms'. When dead and dried out they become hard and brittle, but in life most are soft and flexible. A range of species can be found in British waters, some in rockpools, although you are more likely to find them dead in the strandline.

The remains of these animals turn up in the strandline regularly, though often in a damaged condition. Finding them alive is rather more difficult. Those living in shallow water have a somewhat flattened shape – in deep water they are more spherical.

How to find

■ **Timing** Look for these animals in rockpools at very low tides.
■ **Habitat** Most echinoderms live in deep water, but small sea stars in particular can be quite common in rockpools.
■ **Search tips** Sea stars can be easy to spot in the rockpool, as their tough spiny skins afford good protection against predators. Look very carefully under rocks for brittlestars, which even in life are extremely fragile. Damaged but still spiny remains of sea urchins left atop coastal rocks may indicate that Otters are around.

WATCHING TIPS

These animals are not typically very active, but they are most interesting structurally. In rockpools you may well find Cushion Stars, plump and yellow with five short arms, and in the far South-West the related, recent colonist *Asterina phylactica*, a colourful species with a red inner star marking on its spotted, green skin. Brittlestars are extremely pretty with their long, feathered arms. Avoid handling them, but if you inadvertently damage one you can rest assured that any broken arms will grow back.

Rockpooling

We are lucky in Britain to have so many miles of coastline, much of which is rocky. Sandy and shingle shores offer their own interest, of course, but there is little to rival a really big and diverse rockpool. With a time and a little know-how, you could find a huge diversity of animals.

The best pool

Almost every rockpool will hold something of interest but there are several criteria to consider when seeking a good one with the highest variety of species.

Geography In general, western coasts facing the open Atlantic are best.

Position on the beach The closer to the low-tide mark, the better. Those further up the beach are exposed to the air and sun for longer each day so the water is more likely to evaporate. Further down the beach also means there is more time each day for exciting new arrivals to turn up, during the hours when the open sea covers the rocks.

Size Big is better than small, as there is more room for plants and animals, so more chance that there will be enough food for the predators and scavengers. An enormous rockpool that is big and deep enough to snorkel in should hold a great abundance of wildlife.

Surrounding area If your rockpool is one of many in an extensive stretch of rocky shore, there is likely to be more varied wildlife than if it is part of a small rocky outcrop surrounded by extensive flat sand or shingle.

Discovering rockpool wildlife is a great adventure and a must when on a seaside holiday.

What you'll need

You can have a great day of rockpooling armed with nothing but a bucket – even the bucket isn't strictly necessary. For a more methodical approach, take some collection trays or buckets (transparent or white is best), plus some smaller, clear pots to examine smaller animals – a hand lens is also useful. You might want

Encourage children to **be careful and gentle when using a net** – many rockpool animals are very fragile.

to have a net on standby, but using a net is not always recommended as they can do considerable damage to seaweeds and delicate animals. Use your hands instead – some animals will inevitably escape from you, but you will be extra careful with any that you do handle.

Rockpooling can be dangerous, so some precautions are necessary. Wear very grippy and waterproof footwear for climbing on rocks and avoid dangerously high or steep climbs. Be aware at all times of the state of the tide. It is easy to be cut off from behind if you are on a high rock, so remain alert, and have your phone with you in case you do have problems. Also, avoid handling any animal that looks as though it could bite or sting you. With crabs, you are usually safe from the pincers if you take hold of them by the sides of their carapace.

Welfare

It is essential you are careful and respectful of the animals you are studying; impress this point on your children if you are doing this as a family activity. When walking on rocks, try to avoid crushing molluscs, and when moving rocks around in a pool, do so slowly and gently, and put them back in the same spot afterwards. Don't constantly chase an animal that is trying to get away from you. Handle all your catches very gently, if at all – sessile and fragile animals like sea anemones should not be touched at all. And if you have animals in collection containers, change the water often, especially on a hot day.

Seaweed in rockpools provides hiding places for everything from fish and molluscs to crabs.

Be safe around rockpools: watch your step, watch the tide, and – most importantly – watch your children!

Strandline finds

Mermaid's Purse

1	2	3	4	5

J F M A M J J A S O N D

Portuguese Man O'War

Beachcombing is excellent fun and warrants a whole book in its own right. When you go beachcombing, you'll want to focus on the strandline: the most recent high-tide mark where marine debris is washed up, often in a relatively narrow band. On some tourist beaches, a mechanical cleaning vehicle regularly goes over the sand or shingle, destroying the strandline. Such beaches are really no good at all for beachcombing. Head for a quiet beach on a west-facing shore for the highest chance of discovering something out of the ordinary.

Typically items found on the strandline are of local origin, but could include objects carried thousands of miles across the world by ocean currents. Portuguese Man O'Wars are found in oceans all around the world and may wash up in large numbers after heavy winter storms.

How to find

■ **Timing** Visit at low tide, and if possible pick a time after there has been some heavy weather at sea, as this increases the chances of unusual items being washed up. If you arrive very early or late in the day, you could see land mammals such as Foxes or Hedgehogs visiting the strandline to search for food.

■ **Habitat** All strandlines offer the chance of interesting finds, but shores alongside deep and biologically rich waters will be best.

■ **Search tips** Walk along next to the strandline rather than on it, looking out for anything unusual – also scan other parts of the beach from time to time. Strandline debris can be unpleasantly smelly so you may want to wear gloves if you decide to have a rummage, and it's wise to carry antibacterial gel.

WATCHING TIPS

Among the things you could find are the egg cases of sharks and dogfish ('mermaids' purses'), empty mollusc shells of all kinds, the remains of marine fish and seabirds with attendant scavengers, the hard but unspined 'tests' of sea urchins, and chalky cuttlefish bones. More unusual finds include Goose Barnacles, which resemble large pearl-coloured mussels and arrive in clusters attached to floating objects, and By-the-wind Sailors, jellyfish-like deep blue animals with tentacles and a translucent 'sail'. Sea storms can also bring the similar but larger Portuguese Man o'War ashore – its tentacles can still deliver venomous stings after death so don't touch!

Glossary

Although every effort has been made to avoid excessive jargon and technical terminology in this book, it has been necessary to use certain more specialised terms at times.

Bivalve A mollusc with paired shells that can be opened or closed. Includes clams, mussels, scallops, oysters and cockles.

Blow The breath of a cetacean, appearing as a steamy jet as the animal surfaces and breathes. Its appearance can help with identification.

Costa The leading edge of a dragonfly's wing.

Courtship A series of ritualised behaviours between male and female that occur before mating.

Diurnal Active by day.

Emergence Usually describes the appearance of fully adult insects as they emerge from their pupae, or complete their final moult.

Exuviae Moulted skins of insects.

Foodplant The plant on which immature insect stages feed. Often a species uses a single foodplant species.

Gastropod A mollusc with a single spiral shell – includes land snails and marine species such as whelks and winkles.

Genus A taxonomic grouping of animals, comprising a collection of closely related species, for example, Aeshna (the hawker dragonflies) or Rhinolophus (the horseshoe bats).

Herps A term meaning 'amphibians and reptiles', abbreviated from herpetology – the study of these animals.

Hibernaculum A place where reptiles and amphibians hibernate – can be natural or artificial (for example a sheet of corrugated metal).

Hibernation Passing the winter months in a deeply inactive state – a survival strategy for many animals.

Hymenoptera Bees, wasps, ants and sawflies.

Larva An immature insect – usually used for those species that pass through a pupa stage, such as flies and beetles.

Lepidoptera Butterflies and moths.

Mandibles Biting mouthparts on an insect.

Moult Shedding of the skin to allow growth – occurs in reptiles and many invertebrates.

Mouthparts The varied structures on an insect or other invertebrate's head that are used for feeding, such as the sucking proboscis of butterflies or biting mandibles of beetles.

Nocturnal Active at night.

Nymph An immature insect – usually used for those species that do not pass through a pupa stage but moult directly from a final immature stage to a full adult, such as grasshoppers and dragonflies.

Odonata Dragonflies and damselflies.

Order A taxonomic grouping of animals – for example, Rodentia (rodents) or Lepidoptera (butterflies and moths). It is one level below class (e.g. mammals, insects) and one above family (e.g. squirrels, hawkmoths).

Orthoptera Grasshoppers, crickets and related species.

Ovipositor The egg-laying tube of a female insect – often a prominent anatomical feature.

Palmate Describing webbing between digits on a foot, also the flattened, hand-like appearance of Fallow Deer antlers.

Pedipalps Part of the mouthparts of arachnids, sometimes modified for sensing or grabbing (often referred to as pincers or palps).

Pheromones Chemicals released by one sex of an animal to attract the opposite sex.

Pruinescence A dusting of a blue-grey wax-like substance, giving a blueish appearance. Seen in mature males of some dragonflies and damselflies.

Pterostigmata A cell in the outer wing of insects, like dragonflies, which is often thickened or coloured so stands out from other cells.

Pupa A life stage of some insects, during which a larva transforms into an adult, within a protective shell (which may or may not be enclosed in a cocoon made of silk).

Rut The annual mating season in deer, when males compete through antler-clashing battles in order to attract females.

Scent brand A dark marking on a male butterfly's wing, formed from specialised, odour-producing scales which have a role in courtship.

Sequester To store toxic chemical compounds from the diet inside body tissues – a defence mechanism used by many insect species.

Sessile Describes a non-moving animal – for example, a barnacle, which is fixed to its spot on a rock.

Subimago A stage in the development of some insects (such as mayflies) between the nymph and imago in which the insect is able to fly but becomes mature only after a further molt.

Sward Ground layer of vegetation, usually grass, in the context of its height.

Test The hard exoskeleton of a sea urchin.

Further reading and resources

Books

Top-quality field guides to even the most obscure animal groups are now available. In particular, British Wildlife Publishing (now an imprint of Bloomsbury Publishing) has produced superb field guides to some of the more challenging invertebrate groups, including bees and micro-moths, and WILDGuides publishes field guides covering mammals, reptiles and amphibians, freshwater fishes, hoverflies, spiders and more. The *RSPB Handbook of the Seashore* by Maya Plass (Bloomsbury Publishing) is an invaluable guide to our coastal and rockpool wildlife, while *The Essential Guide to Beachcombing and the Strandline* by Steve Trewhella and Julie Hatcher (Wild Nature Press) should come with you on every beach and seaside walk.

Websites

Every category of creature attracts its own crowd of enthusiasts – some more than others, of course. Here are some useful online resources for general wildlife watching and for a range of particular animal groups.

GENERAL

The RSPB:
rspb.org.uk

The Wildlife Trusts:
wildlifetrusts.org

The National Trust:
nationaltrust.org.uk

The Canal and River Trust:
canalrivertrust.org.uk

iRecord (a national scheme for recording wildlife sightings):
brc.ac.uk/irecord

The Countryside Code:
gov.uk/government/publications/the-countryside-code/the-countryside-code

MAMMALS

Mammal Society:
mammal.org.uk

Bat Conservation Trust:
bats.org.uk

British Marine Life Study Society (cetaceans):
glaucus.org.uk/Cetaceaz

REPTILES AND AMPHIBIANS

Froglife:
froglife.org

Reptiles and Amphibians of the UK:
herpetofauna.co.uk

FRESHWATER FISH

Fish UK:
fish-uk.com

INSECTS (GENERAL)

Buglife:
buglife.org.uk

BUTTERFLIES

Butterfly Conservation:
butterfly-conservation.org

UK Butterflies:
ukbutterflies.co.uk

MOTHS

UK Moths:
ukmoths.org.uk

ODONATA

British Dragonfly Society:
british-dragonflies.org.uk

BEES, WASPS AND RELATIVES

Bumblebee Conservation Trust:
bumblebeeconservation.org

BEETLES

UK Beetle Recording:
coleoptera.org.uk

TRUE BUGS

British Bugs:
britishbugs.org.uk

SPIDERS

British Arachnological Society:
britishspiders.org.uk

Spider and Harvestman Recording Scheme:
srs.britishspiders.org.uk

Many other **useful links** are given at appropriate points in the text.

Index

Photograph credits

Bloomsbury Publishing would like to thank the following for providing photographs and for permission to reproduce copyright material within this book. While every effort has been made to trace and acknowledge all copyright holders, we would like to apologise for any errors or omissions and invite readers to inform us so that we can make corrections to future editions..

Abbreviated photo sources: GI = Getty Images; MT = Marianne Taylor; NPL = Nature Photo Library; RSPB = RSPB Images; SS = Shutterstock.

Cover photos, front: main, Mark Hamblin/RSPB; bottom, left to right Richard Revels/RSPB; Education Images/GI; David Tipling/RSPB; **spine**: David Kjaer/RSPB; **back**, left to right: Drew Buckley/RSPB; Paul Kay/GI; Laurie Campbell/RSPB.

Inside photos, 1 RTImages/SS; **4–5** both MT; **8L** MT; **8R** Nancy Bauer/SS; **9L** Abi Warner/SS; **9R** Menno Schaefer; **10L** Erni/SS; **10R** Keith Burdett/Alamy; **11L** Anatoly Kovtun/SS; **11R** Teele Engaste/SS; **12L** Rudmer Swerver/SS; **12M** Jesus Giraldo Gutierrez/SS; **12R** Erni/SS; **13L** Paul Tymon/SS; **13R** Erni/SS; **14L** Rudmer Zwerver/SS; **14M** David Chapman/Alamy; **14R** Survivalphotos/Alamy; **15L** Jenny Cottingham; **15R** Ian Schofield; **16L** Erni/SS; **16R** Colin Robert Varndell/SS; **17** both Miroslav Klavko; **18L** Taavo Kuusiku/SS; **18R** Erni/SS; **19T** Erni/SS; **19B** BL Footage/SS; **20L** Martin Mecnarowski/SS; **20R** Igor Grochev; **21L** Erni/SS; **21R** MT; **22L** MT; **22R** Ian Rentoul/SS; **23L** Sandra Standbridge/SS; **23R** Peter Wey/SS; **24L** Soru Epotok; **24R** Jackal Photography/SS; **25** both Erni/SS; **26L** Giedriius/SS; **26TR** MT; **26BR** Margaret Welby/Alamy; **27** both Erni/SS; **28L** Erni/SS; **28R** Peter Turner Photography; **29L** Erni/SS; **29R** Belizar/SS; **30L** Erni/SS; **30R** Roman Kurdo/SS; **31L** Stephan Morris/SS; **31R** Glyn Thomas/Alamy; **32L** Erni/SS; **32R** MT; **33L** Stephan Morris/SS; **33R** Philip Bird LRPS CPAGB/SS; **34L** DaveMHunt Photography/SS; **34R** Mark Caunt/SS; **35L** MT; **35R** Erni/SS; **36** both Paul Reeves Photography/SS; **37L** Erni/SS; **37R** FLPA/Alamy; **38** both MT; **39L** DaveMHunt Photography/SS;; **39R** Ian Rentoul/SS; **40** both MT; **41L** Andrew M Allport/SS; **41R** MT; **42L** Erni/SS; **42R** Rudmer Zwerver/SS; **43L** Erni/SS; **43R** Rudmer Zwerver/SS; **44L** David Dohnal/SS; **44R** Bwana Mkubwa/SS; **45L** David Dohnal/SS; **45R** Laurie Campbell/NPL; **46L** Juan Gracia/SS; **46R** Geza Farkas/SS; **47L** Tory Kallman/SS; **47R** MT; **48BL** Will Gray/GI; **48TR** David Havel/SS; **49T** Photo Cornwall/Alamy; **49B** Lindsey Parnaby/Stringer/GI; **50L** Mark Carwardine/NPL; **50M** Anthony Pierce/Alamy; **50R** Tory Kallman/SS; **51L** Heiti Paves/Alamy; **51R** John D McHugh/Staff/GI; **52L** Elise V/SS; **52R** Colette3/SS; **53** both Alessandro De Maddalena/SS; **54L** Richard Darby/SS; **54R** Dave McAleavy/SS; **55L** Gudkov Andrey/SS; **55R**

Kamonrat/SS; **56L** Stephen Farhall/SS; **56R** Economica20/SS; **57** both Bernd Woltr/SS; **58T** Steve McWilliam/SS; **58BL** Rob Kemp/SS; **58BR** imageBroker/Alamy; **59T** Barsan Attila/SS; **59B** blickwinkel/Alamy; **60L** Geza Farkas/SS; **60R** Real PIX/SS; **61L** ; **61R** Gucio_55/SS; **62L** David Dohnal/SS; **62R** Geza Farkas/SS; **63L** MT; **63R** Taviphoto/SS; **64L** Colin Robert Varndell/SS; **64R** Dimitrios Vlassis/SS; **65L** Stephen Morris/SS; **65R** Pollywog/Alamy; **66L** MT; **66R** Abi Warner/SS; **67** both Martin Fowler/SS; **68** both Erni/SS; **69L** Erni/SS; **69R** Tiberiu Sahlean/SS; **70L** Hector Ruiz Villar/SS; **70R** Jack Perks/SS; **71L** Rudmer Swerver/SS; **71R** Tiberiu Sahlean/SS; **72L** Neil Burton/SS; **72R** Erni/SS; **73L** Colin Robert Varndell/SS; **73R** Erni/SS; **74** both John Navajo/SS; **75** both MT; **76T** Bluecrayola/SS; **76B** Joe Blossom/Alamy; **77T** Colin Robert Varndell/SS; **77B** Kay Roxby/SS; **78L** Kletr/SS; **78R** Rostislav Stefanek/SS; **79L** Krasowit/SS; **79R** Mark Caunt/SS; **80L** Rostislav Stefanek/SS; **80R** a-plus image bank/Alamy; **81L** Jack Perks/SS; **81R** Rudmer Zwerver/SS; **82** both Vladimir Wrangel/SS; **83L** Trybex/SS; **83R** Erni/SS; **84L** Vladimir Wrangel/SS; **84R** Erni/SS; **85L** R. Maximiliane/SS; **85R** Erni/SS; **86L** FLPA/Alamy; **86R** Becky Gill/SS; **87L** Sá/Nature Picture Library/GI;; **87R** digitalunderwater.com/Alamy; **88L** Mohamed Tazi Cherti/SS; **88R** Natureworld/Alamy; **89** both MT; **90L** MT; **90R** Tony Mills/SS; **91L** Jerome Whittingham/SS; **91R** MT; **92L** Philp Bird LRPS CPAGB; **92R** MT; **93** both MT; **94L** Marek Miersejewski/SS; **94R** Simon Kovacic/SS; **95L** Bernd Wolter/SS; **95R** Arie v. d. Wolde/SS; **96L** Sandra Standbridge/SS; **96R** Maciej Olszewski/SS; **97L** Alslutsky/SS; **97R** Roger Meerts/SS; **98L** Sandra Standbridge/SS; **98R** Shutterschock/SS; **99L** Lourdes Photography/SS; **99R** Sandra Standbridge/SS; **100L** Simon Kovacic/SS; **100R** MT; **101L** SadlerC1/SS; **101R** Getty Visuals Unlimited, Inc./Robert Pickett/GI; **102** both MT; **103L** Ivan Marjanovic/SS; **103R** Larry Doherty/Alamy; **104L** Ger Bosma Photos/SS; **104R** Martin Fowler/SS; **105L** Steve McWilliam/SS; **105R** Andi111/SS; **106L** Jiri Prochazka/SS; **106R** Florian Teodor/SS; **107L** Martin Fowler/SS; **107R** Shutterschock/SS; **108L** MT; **108TR** Eileen Kumpf/SS; **108BR** Rudmer Zwerver/SS; **109L** Colin Rober Varndell/SS; **109R** Shutterschock/SS; **110L** MT; **110R** IanRedding/SS; **111L** Sandra Standbridge/SS; **111R** Lourdes Photography/SS; **112L** Timelynx/SS; **112R** Sandra Standbridge/SS; **113L** Pieter Bruin/SS; **113R** Sandra Standbridge/SS; **114L** MMCez/SS; **114R** TTstudio/SS; **115L** Sandra Standbridge/SS; **115R** Nature Photographers Ltd/Alamy; **116L** Sandra Standbridge/SS; **116R** Shutterschock/SS; **117** both Chris Button/SS; **118** both MT; **119L** Martin Fowler/SS; **119R** MT; **120L** MT; **120R** Our Wild Life Photography/Alamy; **121L** Martin Fowler/SS; **121R** Nik Bruining/SS; **122L** Colin Robert Varndell/SS; **122R** Rudmer Zwerver/SS; **123L** JPS/SS; **123R** Steve Taylor ARPS/Alamy; **124** both MT; **125** both

MT; **126L** Marek Mierzejewski/SS; **126R** Sandra Standbridge/SS; **127L** Rasmus Holmboe Dahl/SS; **127R** Colin Robert Varndell/SS; **128** both MT; **129L** Mr Meijer/SS; **129TR** Ivan Marjanovic/SS; **129BR** JohnatAPW/SS; **130L** Martin Fowler/SS; **130R** Erik Karits/SS; **131L** Jacques Vanni/SS; **131R** Gubernat/SS; **132L** Marek R. Swadzba/SS; **132R** Ivan Marjanovic/SS; **133L** IanRedding/SS; **133R** Sandra Standbridge/SS; **134L** Chekaramit/SS; **134R** Marek R. Swadzba/SS; **135L** AABeele/SS; **135R** Dmitry Fch/SS; **136BL** blickwinkel/Alamy; **136TR** Nigel Cattlin/FLPA; **137T** Roger Tidman/FLPA; **137B** Paul Glendell/Alamy; **138L** Martin Fowler/SS; **138R** Sandra Standbridge/SS; **139L** Herik Larsson/SS; **139R** Michal Hykel/SS; **140L** Gucio_55/SS; **140R** Gucio_55/SS; **141L** Chekaramit/SS; **141R** Neil Hardwick/Alamy; **142L** Jussi Lindberg/SS; **142R** Mark Anthony Ray/SS; **143L** ajt/SS; **143R** Neil Hardwick/SS; **144L** Cristian Gusa/SS; **144R** Marek R. Swadzba/SS; **145L** Besjunior/SS; **145TR** Martin Fowler/SS; **145BR** Martin Fowler/SS; **146L** Simon Kovacic/SS; **146R** Bildagentur Zoomar GmbH/SS; **147** both Steve McWilliam/SS; **148L** Tony Mills/SS; **148R** Mircea Costina/SS; **149L** IanRedding/SS; **149R** Guillermo Guerao Serra/SS; **150L** Martin Fowler/SS; **150R** IanRedding/SS; **151L** Henrik Larsson/SS; **151R** Korovin Aleksandr/SS; **152L** IanRedding/SS; **152R** Irina Kozorog/SS; **153** both Joke Stuuman/NiS/Minden Pictures; **154–155** all MT; **156L** IanRedding/SS; **156R** Neil Hardwick/SS; **157L** Martin Fowler/SS; **157TR** IanRedding/SS;; **157BR** Sandra Standbridge/SS; **158L** and **TR** DJTaylor/SS; **158BR** Rosemarie Kappler/SS; **159L** Rosemarie Kappler/SS; **159TR** Bildagentur Zoonar GmbH/SS; **159BR** Marek R. Swadzba/SS; **160L** Ger Bosma Photos/SS; **160R** Dan Bagur/SS; **161L** IanRedding/SS; **161R** DJTaylor/SS; **162L** Henrik Larsson/SS; **162R** Nigel Downer/SS; **163L** Martin Fowler/SS; **163R** IanRedding/SS; **164L** corlaffra/SS; **164R** CHWeiss/SS; **165L** Henrik Larsson/SS; **165R** Henri Koskinen/SS; **166L** IanRedding/SS; **166R** Neil Hardwick/SS; **167L** Matauw/SS; **167R** DJTaylor/SS; **168L** IanRedding/SS; **168R** Eldie Aaron Justim/SS; **169L** Edwin Butter/SS; **169R** IanRedding/SS; **170TR** Bildagentur Zoonar GmbH/SS; **170ML** Mirko Graul/SS; **170BL** Rasmus Holmboe Dahl/SS; **171T** IanRedding/SS; **171BL** Jacques VANNI/SS; **171BR** pp1/SS; **172L** Ger Bosma Photos/SS; **172R** Martin Fowler/SS; **173L** Mirko Graul/SS; **173R** North Devon Photography/SS; **174L** MT; **174R** Mirko Graul/SS; **175L** Henrik Larsson/SS; **175R** Emjay Smith/SS; **176** both Macgorka/SS; **177L** MT; **177R** Fabio Sacchi/SS; **178L** MT; **178R** Macgorka/SS; **179L** Rudmer Zwerver/SS; **179R** Peter Reijners/SS; **180L** MT; **180R** Macgorka/SS; **181L** MT; **181R** Alslutsky/SS; **182L** Martin Fowler/SS; **182R** MT; **183L** Mirko Graul/SS; **183R** MT; **184L** Marek Mierzejewski/SS; **184R** Johannes Dag Mayer/SS; **185L** shaftinaction/SS; **185R** Ian_Sherriffs/SS; **186L** Peter Eggermann/SS; **186R** Jiri Prochazka/SS; **187L** shaftinaction/SS; **187R** Alslutsky/SS; **188L** jcm32/SS; **188R** Andrew M. Allport/SS; **189L** Torsten Dietrich/SS; **189R** Sandra Standbridge/SS; **190L** Hector Ruiz Villar/SS; **190R** Michal Hykel/SS; **191L** Fabio Sacchi/SS; **191R** Tony Mills/SS; **192L** Marco Maggesi/SS; **192R** Martin Fowler/SS; **193L** Martin Fowler/SS; **193R** Maciej Olszewski/SS; **194L** MT; **194R** IanRedding/SS; **195L** Pavel Hajer/SS; **195R** MT; **196L** Ainars Aunins/SS; **196R** ethylalkohol/SS; **197** both Fabio Sacchi/SS; **198L** TTStudio/SS; **198R** Florian Teodor/SS; **199** both MT; **200** both MT; **201L** Hansh/SS; **201R** MT; **202L** MT; **202R** Alslutsky/SS; **203L** Pavel Hajer/SS; **203R** David J. Martin/SS; **204L** Mark Carthy/SS; **204R** Colin Robert Varndell/SS; **205L** Martin Fowler/SS; **205R** Fabio Sacchi/SS; **206L** Fabio Sacchi/SS;

206R Paul Reeves Photography/SS; **207L** A.G.A./SS; **207R** PJ Photography/SS; **208L** Magdalenawd/SS; **208L and R** IanRedding/SS; **209L** Nick Upton/Alamy; **209M** Irina Orlova/SS; **209R** FLPA/Alamy; **210L** Martin Fowler/SS; **210R** Henrik Larsson/SS; **211L** Ant Cooper/SS; **211R** Timelynx/SS; **212L** Ed Phillips/SS; **212R** Sue Robinson/SS; **213L** blickwinkel/Alamy; **213M** BIOSPHOTO/Alamy; **213R** NPL/Alamy; **214T** Linda George/SS; **214B** Timelynx/SS; **215** both Peter Turner Photography/SS; **215B** Ulf Wittrock/SS; **216L** Cristian Gusa/SS; **216R** Ernie Janes/SS; **217L** Henrik Larsson/SS; **217R** shaftinaction/SS; **218L** Martin Fowler/SS; **218R** IanRedding/SS; **219L** Sebastian Janicki/SS; **219R** Marek Velechovsky/SS; **220L** IanRedding/SS; **220R** Alex Puddephatt/SS; **221L** HHelene/SS; **221R** Nigel Cattlin/Alamy; **222L** IanRedding/SS; **222R** Martin Fowler/SS; **223L** IanRedding/SS; **223R** Martin Fowler/SS; **224L** Ant Cooper/SS; **224R** Hamik/SS; **225L** Ant Cooper/SS; **225R** Martin Fowler/SS; **226L** Denis Vesely/SS; **226R** Martin Fowler/SS; **227L** Montypeter/SS; **227R** IanRedding/SS; **228L** MR Aukid Phumsirichat/SS; **228R** Bildagentur Zoonar GmbH/SS; **229L** A.S.Floro/SS; **229R** IanRedding/SS; **230** both Tommy Lee Walker/SS; **231L** Martin Fowler/SS; **231R** shaftinaction/SS; **232L** Jiri Prochazka/SS; **232R** IanRedding/SS; **233L** NHJ/SS; **233R** Manuel Overwien/SS; **234T** NPL/Alamy; **234B** FLPA/Alamy; **235T** Natural History Museum/Alamy; **235B** Jack Perks/Alamy; **236L** Martin Pelanek/SS; **236R** Robert Shantz/Alamy; **237L** Maciej Olszewski/SS; **237R** Henrik Larsson/SS; **238L** Christian Musat/SS; **238R** nechaevkon/SS; **239L** Arto Hakola/SS; **239R** Martin Fowler/SS; **240L** Martin Pelanek/SS; **240R** Henrik Larsson/SS; **241L** Dan Bagur/SS; **241R** Maciej Olszewski/SS; **242L** Ger Bosma Photos/SS; **242R** aaltair/SS; **243L** Ian Redding/SS; **243R** Martin Fowler/SS; **244L** Ian Redding/SS; **244R** yod67/SS; **245L** Martin Fowler/SS; **245R** MiQ/SS; **246L** Torsten Dietrich/SS; **246R** lkpro/SS; **247L** David Peter Ryan/SS; **247R** Jiri Prochazka/SS; **248L** Chris Moody/SS; **248R** Sarah2/SS; **249L** IanRedding/SS; **249R** Marco Uliana/SS; **250T** Monkey Business Images/SS; **250B** Jason Smalley Photography/Alamy; **251T** Pat Tuson/Alamy; **251B** MintImages/SS; **252L** Jiri Prochazka/SS; **252R** Krasowit/SS; **253L** Stephen Dalton/Minden Pictures/GI; **253R** Premaphotos/SS; **254L** Nigel Downer/SS; **254R** Christian Musat/SS; **255L** nujames10/SS; **255R** Pavel Krasensky/SS; **256L** Tony Mills/SS; **256R** Martin Fowler/SS; **257L** Dan Bagur/SS; **257R** KKindl/SS; **258L** Muhammand Naaim/SS; **258R** IanRedding/SS; **259L** Frank Hecker/Alamy; **259R** Anton Kozyrev/SS; **260L** stevie_uk/SS; **260R** thatmacroguy/SS; **261L** Sandra Standbridge/SS; **261R** Colin Robert Varndell/SS; **262** both Dave Pressland/FLPA; **263** both Stephen Dalton/Minden Pictures/GI; **264** both thatmacroguy/SS; **265L** IanRedding/SS; **265R** Jaromir Grich/SS; **266L** Andrew Balcombe/SS; **266R** IanRedding/SS; **267** both Marco Maggesi/SS; **268L** Dan Bagur/SS; **268R** Lynsey Allen/SS; **269L** Harald Schmidt/SS; **269R** TonLammerts/SS; **270L** IanRedding/SS; **270R** HHelene/SS; **271L** IanRedding/SS; **271R** Steve Trewhella/Alamy; **272L** FLPA/SS; **272R** Arterra Picture Library/Alamy; **273L** Andrey Armyagov/SS; **273R** Will Heap/GI; **274L** ECOSTOCK/GI; **274R** Chris Warham/SS; **275L** Cultura RF/Alexander Semenov/GI; **275R** Mark Webster/GI; **276L** Nick Garbutt/GI; **276R** Norbert Wu/Minden Pictures/GI; **277L** Daniela White Images/GI; **277R** Paul Kay/GI; **278** both Alex Mustard/Nature Picture Library/GI; **279T** robertharding/Alamy; **279M** Alex Mustard/Nature Picture Library/GI; **279B** Betty Finney/Alamy; **280L** Paul Kay/GI; **280R** Matt Cardy/Stringer/GI.